The Backyard
DUCK BOOK

The Backyard DUCK BOOK

For the Love of Ducks

Revised Edition

Nyiri Murtagh

PUBLISHING

National Library of Australia Cataloguing-in-Publication entry

Murtagh, Nyiri.

The backyard duck book : for the love of ducks / by Nyiri Murtagh.

Revised ed.

9780643106512 (pbk.)
9780643106529 (epdf)
9780643106536 (epub)

Includes bibliographical references.

Ducks.
Duck breeds.

636.597

Published by
CSIRO PUBLISHING
36 Gardiner Road, Clayton VIC 3168
Private Bag 10, Clayton South VIC 3169
Australia

Telephone: [+613] 9545 8555
Local call: 1300 788 000 (Australia only)
Fax: +61 3 9662 7555
Email: csiropublishing@csiro.au
Web site: www.publishing.csiro.au

Front cover: White Muscovy duck with brood (left); Best in show Muscovy (right)
Back cover: Mallard male (left); Saxony male (right)

Photographs by Nyiri Murtagh, unless otherwise stated.
Illustrations by Rain Hart, unless otherwise stated.
Colour photographs by Rupert Stephenson.

Set in 10/16 Palatino
Edited by Lachlan Garland
Cover and text design by James Kelly
Typeset by James Kelly
Printed by Ingram Lightning Source

Feb26_RP_ILS

Foreword

As a breeder and exhibitor of ducks for 50 years, I would recommend this book to anyone with an interest in ducks and duck husbandry. *The Backyard Duck Book* provides a valuable insight into owning, housing and breeding this most interesting bird.

Drawing on over 10 years involvement in breeding ducks, Nyiri has provided a practical, common-sense guide to keeping, housing and breeding ducks. The many illustrations provide excellent visual examples of what is explained in the text.

Nyiri's knowledge has been gained from extensive research and travels, but most importantly from the advice and guidance of some of Australia's most experienced breeders.

David Green
Executive Member, Waterfowl Fanciers Association of Queensland

Contents

1
Introduction: for the love of ducks

I began watching ducks when in search of solace. I set up their yards like a park, with palm trees, bamboo, gum trees, a large pond and lots of green grass that resembled a lush lawn. I built a pergola from railway sleepers and would often sit quietly with a cup of tea or a good stiff drink, depending on my mood, and watch my birds. This pastime turned into the best stress relief and contemplation time any doctor could have prescribed. It wasn't long before I was admiring these natural clowns. To this day, I remain intrigued by their playfulness and addicted to their beauty and grace.

I don't consider myself a duck expert. I don't show my birds – never have and never will. And I am not a member of any poultry association or club. I simply enjoy observing my birds. Over the past 10 years, they have taught me a great deal. They've given me more laughs and entertainment than I could poke a stick at, and have helped me find the tranquility I was searching for.

I still feel I have a lot to learn, and I have no intention of ceasing to ask questions or experiment. I guess the key to success, no matter what you are interested in, lies in listening; not only to advice, but also to other breeders' opinions and information, which can be very helpful.

It is my hope that one day many of the less common breeds of ducks will be prolific and of a quality that will hold its own alongside any exhibition stock in the show arena. But that begins with people like you and me sharing information that will enhance and improve the breeds. Far too many 'old

Figure 1.1: It pays to put a lot of thought into your ducks' environment.

timers' take their vast knowledge to the grave, which is detrimental to the birds they said they held a passion for!

There are many things to consider when owning ducks: from their enclosure, to their safety at night, swimming facilities, health and nutrition, mating, incubation of eggs, and the breed itself. I hope *The Backyard Duck Book* will be of help to you. It is a compilation of my own experience over the very short 10 years that I have been breeding and enjoying ducks.

Of course the first thing you must do is ask yourself a few questions. Why do you want a duck? What are your intentions? What can you give a duck? A duck is not like a hen, or a dog or cat for that matter.

I am of the belief that it costs the same to feed a mongrel duck as a good one, and at the end of the day, the good one is still worth more to you than the mongrel. So, if you are going to breed a duck, breed a good one. I take a great deal of enjoyment from admiring the birds that are true representatives of their breeds and I'm sure you will too.

But first you must love the duck for what it is, not for what it might bring. I believe you must always place your ducks first. Think about their environment. Look at the enclosure where you plan to place your ducks and ask yourself: 'Would I be happy to sit here for the better part of the day?'

2
Duck anatomy

Parts of a duck

Novices attending poultry shows may be baffled by terms used by judges or breeders when discussing a duck, or ducks in general. Here is a quick reference to the parts and terms used.

If you become an exhibitor of waterfowl, you will learn that each bird has the potential to earn 100 show points. Each part of a duck is worth a certain number of points.

Figure 2.1: The parts of a duck and the show terms used to describe them.

The duck digestive system

(1) **Bill:** used for shovelling food into the mouth. Swallowing occurs by a natural reflex as the moistened food nears the back of the tongue.

(2) **Oesophagus:** a muscular tube that joins the back of the mouth and the crop. Waves of muscular movements move the food along it and into the crop.

(3) **Crop:** a thin-walled, bag-like structure found near the front and the base of the neck. It is quite prominent after the bird has eaten. This organ acts as a temporary storage for the food, which, during its stay, is softened by the moistening action of the saliva.

(4) **True stomach (proventriculus):** the food passes from the crop through the breast past the wishbone to the true stomach where acidic gastric juice further softens and commences the digestion of proteins.

(5) **Gizzard (also known as the 'grinder'):** performs the function that teeth do in many other species. It is a muscular, thick-walled organ with a tough, leathery inner surface. While the duck is foraging for food, it picks up stones and grit that stay mixed with the food in the gizzard. The combination of the gizzard's muscular movements, its leathery surface and ingested grit breaks down the food to a fine state.

(6) **Duodenal loop (pancreas):** when the food enters here it is treated with secretions from the pancreas, which is located in the loop. There are two parts to the pancreatic juice:

- (a) a component to neutralise the acidic effects of the gastric juice
- (b) a solution of food-splitting enzymes that break up the large food particles into smaller, simpler ones that can be easily absorbed through the intestinal wall and into the blood stream.

(7) **Small intestine:** the principle role of this organ is food absorption. Its structure provides a large absorbing surface of relatively small volume. Features include its length, coiled shape and internal surface, which contains projections known as 'villi'. Food constituents are absorbed through these thin walls into the blood stream and transported to the liver where they are transformed into other forms, for example, amino acids and proteins, and temporarily stored until directed for use in other parts of the bird's body.

(8) Caeca: residual food material passes two blind tubes, called the caeca, whose function is unclear. There is some suggestion that further food breakdown may occur due to microbial action.

(9) Large intestine: this is a water-reabsorbing organ that turns the remaining undigested food into a semi-solid mass. The large intestine is not a particularly long organ in the duck in comparison with the large intestines of other animals.

(10) Rectum: the last part of the large intestine where undigested food (faeces) is temporarily stored prior to evacuation.

(11) Cloaca: a common opening for the digestive, urinary and genital systems of the duck. Kidney waste is added to the faeces at this point.

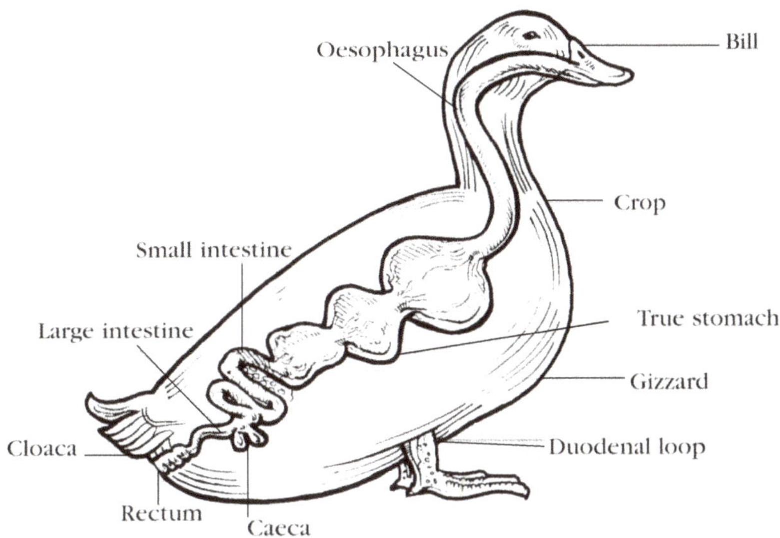

Figure 2.2: The digestive system of a duck.

The duck reproductive system

Female birds have sexual organs on the left side of their bodies. The reproductive system consists of an ovary and an oviduct.

The ovary looks like a cluster of little grapes and contains 1000–1600 egg cells. These cells are extremely small to begin with, but, under hormonal stimulation, some start to accumulate yolk and become enlarged. They are then termed 'ovarian follicles' or 'ova'.

The male has a pair of testes attached to the kidneys and located close to the back of the bird. Semen passes down a tube to a small sac adjacent to the cloaca. A drake's penis is upwards of 5 cm in length, depending on the breed. It is spiral in shape and appears to be serrated.

Egg fertilisation takes place once the drake has inserted his penis into the duck's cloaca and ejected sperm. The sperm swim along the mucous-covered lining of the oviduct until they reach the neck of the oviduct, which is near the ovary. Here the ovum, or female cell, is fertilised as it passes, with yolk attached, into the oviduct. It takes time for sperm to swim into the oviduct, and it can be 7–10 days after mating before fertile eggs are produced.

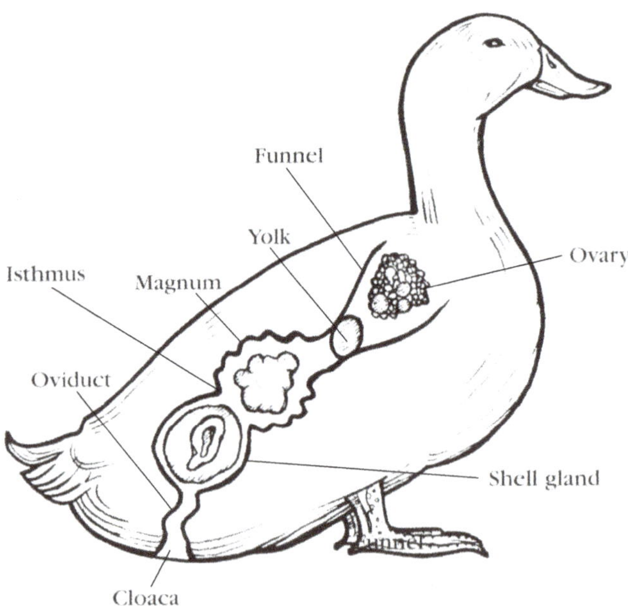

Figure 2.3: The female duck reproductive system. When an egg is laid the vaginal wall is pushed through the cloaca and outside to protect the egg from contact with excreta.

Duck egg formation

There are four stages in the development of an egg:

(1) yolk formation

(2) egg white formation

(3) membrane formation

(4) shell formation.

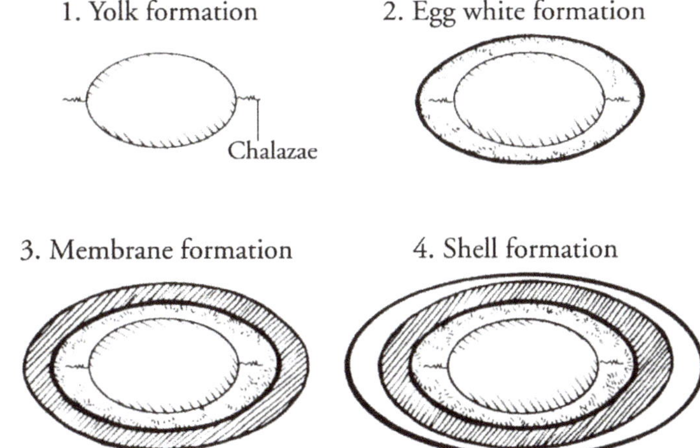

Figure 2.4: Developmental stages of a fertile egg.

Yolk formation

Yolk formation is the first stage in egg development. It takes place in the ovary, where the female sex cell is surrounded by a large amount of fat and protein, and covered with a yolk (vitelline) membrane. The yolk is then released from the ovary into the funnel of the oviduct where, if sperm are present, the egg will be fertilised. The chalazae (twisted strings by which the yolk bag is held in position) are also added at this stage. The egg remains here for 15 minutes before moving to the magnum section of the oviduct.

Egg white formation

During the next 3 hours, the egg white is placed around the yolk as it passes through the magnum. The white of an egg acts as an antibiotic for the duckling embryo growing inside. It has a similar purpose to the colostrum a newborn mammal obtains from its mother's milk for a period after birth.

Membrane formation

During the following hour, an inner and an outer shell membrane form around the white-clad yolk as it passes through the isthmus.

Shell formation

By far the greatest amount of time, some 21 hours, is spent in the gland where the shell is formed around the membranes covering the white and yolk. The egg is now ready for laying, large end first, tapering to the smaller end as it is expelled from the oviduct. The egg is wet when first laid and the shell becomes more brittle as it dries.

Saxony, male – 1st.

Best in show – Champion waterfowl. Pekin.

Best in show – Muscovy.

Best in show – Aylesbury, male.

Best light duck – Welsh Harlequin, male.

Best of breed – Black Muscovy, male.

Black East Indian.

Indian Runner-Saxony, female.

Best of breed – Muscovy.

Best of breed – Cayuga.

Best of breed – Crested Duck, male.

Best of breed – Rouen Claire, female.

Khaki Campbell, female.

Khaki Campbell, male – 1st.

Best of breed – Indian Runner.

Best of breed – Buff Orpington.

Mallard, male.

Best of breed – Saxony, female.

Blue Swedish, female – 1st.

White Muscovy duck with brood.

3
Duck breeds and their standards

With the exception of the Muscovy, all ducks originated from the wild Mallard. Most breeds come in standard and bantam sizes.

The information on availability of breeds is based on talking with breeders and my own travels throughout Australia. Unfortunately, most breeders are not affiliated with associations or clubs and tend to swap and buy birds among themselves. Many do not exhibit their ducks. It is not always easy for new breeders to obtain good quality birds.

Appleyard (Silver)

Origin

The Appleyard, also called the Silver Appleyard, is believed to have been developed by Reginald Appleyard during the 1930s at the Priory Waterfowl Farm in England. Reg's aim was to breed the ideal duck. It is thought that he crossed the Mallard with the White Campbell. This makes sense as the Mallard would give the unusual colouring and the Campbell would contribute the longer body and egg-laying qualities. The breed is classed as ornamental, despite the fact that it is a heavy bird that lays huge white eggs. I guess they thought the Appleyard was too attractive to be considered a 'commercial' line of duck.

There is some evidence to suggest that a few Australian breeders noticed Appleyards overseas and liked them very much, but as quarantine laws prohibited them from importing the breed into Australia, they set about developing their own line of the breed.

Description

The Appleyard is one of the most beautiful breeds of duck. It looks very much like the Rouen, being just as highly coloured, but it is slightly smaller, more erect and less placid than the Rouen.

The Appleyard drake, with its beetle-green head, neck and back, silver-white neck ring, and deep claret chest and shoulders, is a particularly attractive bird. The duck is silvery-white with a heavy flecking of fawn on her back. She is an excellent layer, producing large white eggs.

Appleyards are classified as heavy birds; drakes are quick to mature and make good table birds.

I find the miniature or bantam Silver Appleyard more appealing. It should be an exact replica of its larger counterpart in all but size. They are like petite, colourful garden gnomes, but be warned ... they can fly. They

Figure 3.1: Silver Appleyards, female (left) and male.

will go feral in a heartbeat and, short of loading up the shotgun, it's almost impossible to catch them once they have flown away. In fact, one of mine still lives on a dam about a kilometre away with a flock of shell ducks. To keep your ducks from flying off, pinion their wings when they are ducklings, or trim their wings regularly (the latter is the preferred method). Bantam Silver Appleyards are among the best layers of the small breeds.

A friend of mine breeds bantam Silver Appleyards that have hatched guinea fowl chicks for him. The duck looks very amusing foraging for food with a little flock of striped chicks running behind ready to spring on any bug or grub the 'mother' duck finds for them. Then at the end of the day, they nestle beneath her feathers for warmth. Bantam ducks are brilliant mothers and will successfully raise their young, regardless of whether the young have webbed feet or not. It is not uncommon for a bantam Mallard to lay 20 eggs and hatch 20 ducklings.

Standards

The body should be long and large. The drake should have a green head with faint silver markings on eyebrows and cheeks. The neck should be green flecked with silver, with a white ring on the throat that joins at the back. Shoulders and breast should be claret, and the keel, wing coverts and underbody should be silver. The wing bar should be blue. The back and rump should be green-black, the same colour as the head. Tail feathers should be white-tipped. The duck should have a silver-white head. The neck and underbody should be flecked with fawn. There should be a fawn line through the eyes and the tail should be solid fawn. The wing bar should be blue. The eyes should be dark hazel. Both sexes should have a yellow bill and light orange legs.

Serious defects

Undersized, wrong bill colour, wrong wing bar colour, lack of colour on throat.

Current status

Finding a good Appleyard can be a bit of a challenge. In Australia, individual birds vary a great deal from the standard and lack consistency in feathering.

Aylesbury

Origin

Originally called English Whites, Aylesburys are thought to have been bred for the table in the town of Aylesbury, in southern England. The breed was introduced into Australia in the early 1920s, quite possibly by the late W. Scott, and is the oldest exhibition waterfowl on record. At the first poultry show in 1845, a class was made for the Aylesbury.

Description

The Aylesbury is a massive snow-white duck, with a characteristic horizontal carriage and a broad, deep body. Like the Pekin, the Aylesbury is a white meat duck. It dominated the table-duck trade in Britain for many years.

The Aylesbury's plumage is unlike that of any other breed. It is incredibly soft and downy; these ducks are a lot softer than the Pekin to touch and hold. The drakes have two to three curled feathers in their tail.

Figure 3.2: Aylesbury.

The Aylesbury's legs and feet are bright orange, and its bill is pinkish-white, or flesh coloured. I have noticed, however, that if you run these birds in the sun, the bill tends to take on a yellow tinge. If kept in a shaded area, the bill remains pink. Dish bill is a common problem with this breed.

Aylesburys are not nervous ducks and the purebred is a good layer, laying about 100 eggs a year. The eggs are white and the size of a young goose's egg; there is no mistaking the eggs of an Aylesbury for the eggs of any other breed. Egg numbers tend to decrease in hot weather.

They aren't fast moving ducks. Nor are they as vigorous as other breeds, so swimming ponds should be available for them if you are expecting fertile eggs.

Aylesbury ducklings are rather large and bulbous, like little pears, in shape. They take 2 years to fully mature.

There is a utility strain of the Aylesbury, which was developed by crossing the Pekin and the Aylesbury to produce meat birds. Utility Aylesburys have a yellow bill.

Standards

British poultry standards state that the body should be long, broad and very deep, showing good keel. Ducklings should keel up early, within the first 6 to 8 weeks. The back should be straight, almost flat, and the breast should be full and prominent. The keel should be straight from breast to stern. The wings should be strong and carried close to the sides and fairly high, but not touching across the saddle. The tail should be short, only slightly elevated, and composed of stiff feathers. The head should be strong and powerful, with eyes set near the top of the skull. Eyes should be full. The bill should be strong and wedge-shaped. The head and bill should measure 15–20 cm in length and, viewed from the side, the profile should be almost straight from the top of the skull. The neck should be curved and strong. The legs and feet should be short and strong, and the bones thickset to balance a level carriage. The feet should be straight and webbed. The plumage of both sexes should be white. It should be bright and glossy, resembling satin. Legs and feet should be bright orange, and the bill should be pinkish-white.

Serious defects

Plumage any colour other than white. A yellow, short bill or a dish bill. Little or no keel.

Current status

Today the breed is exceptionally rare, as are top quality specimens. The breed's current position is being seriously challenged by strains of Pekin and Indian Runner. I was fortunate enough to obtain a pair from Ken Younge, of Queensland, and then an unrelated trio from Ron Smith, of New South Wales.

Black East Indian

Origin

There is some reference to the Black East Indian duck being developed in the USA. The Black East Indian was included in the first standards book in 1865. The breed has had many names including Buenos Ayres, Labrador, Black Brazilian and Black East Indies. It is thought to be a descendant of the Black Mallard.

Description

The Black East Indian is a small duck that is regarded in Australia as a bantam. Both duck and drake are black, with a bottle-green sheen. Even the bill, legs and webs are black, although the legs and webs tend to become more orange with age.

Black East Indians don't lay a large number of eggs, but they do have a reputation for being good mothers. They are very hardy and are excellent foragers. The first eggs of the laying season are black or dark grey, gradually fading in colour as the laying season progresses.

As Black East Indians age, they develop more and more white feathers. Some strains turn white more quickly than others, so when breeding, select those that keep their colour longest.

Figure 3.3: Black East Indian.

Standards

Carriage should be clear of the ground, lively and alert. The body should be short, neat and broad. The breast should be round and prominent. The head should be neat and round, with a high skull. The bill should be short and the eyes full. The plumage of both sexes should be bottle–green-black, with no white or purple feathers to spoil the sheen. Eyes should be dark and the bill slate-black, with blotches of olive. Legs and webs should be black.

Serious defects

White feathers, wrong colour bill.

Current status

Rare, but available. Some breeders, who really breed Black Mallards, call their ducks Black East Indians because it is illegal to keep Black Mallards.

Blue Swedish

Origin

No one is really sure of the origin of the Blue Swedish breed. A species of blue duck was recorded in the north of England early in the 19th century, and they were thought to have been introduced from Holland. Blue ducks were also shown in England under the name of Orpington and some people suggest that these ducks were renamed Blue Swedish. However, around 1884 the breed was introduced into the USA from Germany. There is some suggestion that the Blue Swedish in Australia are simply a sport of the Appleyard.

Description

Blue Swedish are medium-heavy birds with blue bodies, white bibs and white outer wing primaries. They are hardy and good foragers. Blue Swedish ducks lay blue, grey or tinted eggs.

Figure 3.4: Blue Swedish.

A good Blue Swedish is difficult to breed. It does not breed true in the blue form. Out of a breeding of two blue, only 50% of the ducklings will have the blue-grey colour of their parents. Half of the remainder will be black and white, and the rest silver. However, crossing a silver bird with a black and white will produce 100% blue-grey ducklings.

Bantam Blue Swedish are called Bibbed Ducks. They are bred to the same standard, but have a different name.

Standards

Carriage should be at about 25 degrees. The body should be long. The head should be a green-black with no brown feathering, and the white bib should begin at the front, below the bill and should form a neat circle just above the breast. The feathers covering the body should be blue, laced finely in black. The primaries should be white. The legs should be brownish-orange. The duck's bill colour should be blue, and the drake's green with a black bean.

Serious defects

Bib too large, too small or broken. No white primaries or too many white primaries. White patches throughout body colour, brown feathers.

Current status

Good Blue Swedish are rare.

Campbell (Khaki)

Origin

Adele Campbell, who intended to produce a good laying bird, developed the Campbell at the turn of the 19th century. The breed is the result of crossing strains of Fawn and White Indian Runners, Mallards and Rouens. The breed was introduced into Australia in 1901. Since then it has become increasingly popular as a breeder, egg layer and snailer.

Description

While the plumage of the original Campbell drake followed the colouring of a Mallard male, though lighter and with no white ring around its throat, the duck was the colour of withered grass. Both drake and duck have green bills.

There are three varieties of Campbell: White, Khaki and Dark.

The Khaki Campbell is the variety resulting from the colour selection of the original Campbell. The drake's head, neck, stern and wing bar are a lustrous green-bronze, while the remainder of the body is an even shade of khaki. Its legs should be orange. The duck is a warm khaki throughout, although the neck, head and wings are a slightly darker shade of khaki. Each feather on the rest of the body is narrowly laced with a slightly lighter shade of khaki. Although the feet and legs of the original Campbell ducks were an orange-brown, today the legs and feet should be as near body colour as possible.

The Dark Campbell was bred by H.R.S. Humphrey by crossing Khaki ducks with Pekin drakes. This mating resulted in a variety that, when crossed with pure Khaki ducks, produces offspring that are described as being sex linked. This simply means that female and male ducklings are different colours at hatching, so you can easily pick male from female. In this case, the young drakes are lighter in colour than the young ducks.

Figure 3.5: Khaki Campbell, male.

The Dark Campbell duck is very much darker than the Khaki duck and has tiny pencilling, rather than lacing, through her feathers. The Dark drake, surprisingly, is lighter than the Khaki drake. The general colour is light brown; the feathers being finely pencilled with grey-brown. The head, neck, stern and wing bar are a beetle-green. The drake's feet and legs are bright orange, while those of the duck are nearer body colouring.

Campbells are credited with being the best of the duck breeds for egg production, boasting reports of 360 eggs per year. If you want good egg yields, however, be certain you acquire authentic Campbells, as some crossbreds are being sold as genuine Campbells. Another problem with the available Khakis is that breeders who breed for show are more concerned with how the bird looks, as opposed to the egg production, so look for a good laying strain.

Campbells are extremely hardy and active and will forage over a large area quite happily. Of all breeds of true ducks (Muscovies not being a true duck), the Campbells have the greatest desire to swim.

Standards

Campbells should have a good length in body. They should be alert in their carriage, slightly upright and symmetrical. The head should be carried high, with the shoulders higher than the saddle, and the back showing a gentle slant from shoulder to saddle. The whole carriage should not be too erect, but should not be so low as to cause the bird to waddle. The body should be wide, deep and compact, appearing slightly compressed. Activity and foraging power should be retained, without loss of depth and width of body generally. The wings should be carried closely, and rather high. The tail should be short and small, rising slightly, and the drakes should have the usual curled feathers. The neck should be of medium length, slender and refined, almost erect. The head should look like it fits on the neck. The skull and jaw should be refined, and the face smooth. The bill should be proportionate and of medium length, depth and width. It should be well set, in a straight line with the top of the skull. Eyes should be full, bold and bright, showing alertness and expression. The legs and feet should be of medium length and set well apart to allow good abdominal development. Plumage should be tight and silky, giving a sleek appearance.

Serious defects

Yellow bill. White bib or neck ring on drakes or ducks. Streaks of dark feathers through the eye in ducks. White or light underpart or top of wings in either sex. White in wing bar. Orange legs on the ducks.

Current status

I have found that in Australia this is a very debatable breed, with many people holding different views on what a good Campbell should look like. Even in some of the better-known books you will often find they have at least three different listings and photos for them.

Cayuga

Origin

The Cayuga's native land is North America. The breed is thought to have been developed from Black East Indian ducks in the region of Lake Cayuga, in upstate New York.

Description

The Cayuga is a general-purpose, green-black bird that furnishes a good supply of eggs. In keeping with her ebony plumage, the shells of the duck's first eggs for the season are black or dark grey. The eggs fade to light grey or blue as the season progresses. The ducklings are black, with black feet and bills, and just a hint of yellow in the down on the breast. Cayugas are one of the hardiest and quietest of the duck breeds, and make very good broodies.

Unfortunately many of the most lustrous black Cayugas develop some white feathers over time. This is particularly true of the ducks. Breeders control this to some extent by mating green-black drakes with no trace of white with ducks that have no white feathers as young birds. In checking drakes for traces of white, be sure to inspect under the bill. Pure black ducklings are often obtained from ducks with just a hint of brown under their bills and wings.

Standards

Carriage should be lively and clear of the ground from breast to stern. The body should be long, broad and deep. The breast should be prominent,

Figure 3.6: Cayuga.

with the keel well forward and forming a straight line from stem to stern. The tail should be carried well out and closely folded. The drakes should have two or three well-curled feathers. The head should be large. The bill should be long, wide and flat, and well set in a straight line out from the tip of the eye. The eyes should be full. The neck should be long and tapering, with a graceful curve. The legs and feet should be strong boned and placed midway in the body, giving the bird a similar carriage to that of the Rouen. Both sexes should possess lustrous green-black plumage, free from purple or white. The whole of the back and the upper part of the body, including wings, breast and underbody, should be deep black. A brown or purple tinge is undesirable. In both sexes, the bill should be slate-black, with a dense black saddle in the centre (but not touching the sides). The eyes should also be black. The legs and webs should be dull orange-brown.

Serious defects

Red or white feathers. Orange bill, dished bill.

Current status

Plentiful.

Crested Duck

Origin

Crested Ducks were bred for exhibition in Europe.

Description

The Crested Duck is very striking. The large unusual tuft of feathers on its head resembles a hat from the days when women wore outlandish and striking ornaments on their heads.

This breed is a prolific egg layer and could be placed in many classes, as it would seem they come in all shapes and sizes. Some resemble Pekins, while others look like Blue Swedish or Indian Runners. I once owned a Rouen that had a crest.

I tend to think the Crested are simply crossbred ducks; the crest having been developed and encouraged by waterfowl fanciers for shows. The crested gene is dominant. Thus if you cross a Crested Duck with any other breed, a percentage of the ducklings will have crests, and the remainder will carry the gene for reproducing crests. Hence you can have a Crested Pekin,

Figure 3.7: Crested Duck.

Crested Orpington, Crested Indian Runner and so on; the list is endless. The Crested is classed as a light breed and lays a white egg.

Crested Ducks carry a gene that prevents about 25% of the fertile eggs hatching. An additional 25% do not hatch with the crest.

Standards

Crested Ducks should have a fine head and be semi-upright in carriage. The crest should be on the top of the head, and very large and full. The larger the crest, the more desirable the look. Any colour is acceptable in these birds.

Serious defects

Poor crests. Crest not on top of head.

Current status

Plentiful.

Duclair

Origin

The Duclair, or Rouen Claire as it is also called, originated in northern France. It is more than likely a sport of the Rouen.

Description

A long, large duck, the Duclair is classed as a heavy breed. It looks a lot like the Rouen, although it is a little lighter in colour and has no keel. It is often confused with the Appleyard.

The general shape of this breed is long and graceful, with good width. The most important feature of the breed is its length. The ideal length is 90 cm from the point of the bill to the end of the tail; that is, with its neck extended.

Duclairs are very attractive birds, are quite active and make good-sized garden ducks. They lay white or green eggs.

This breed is the 'next size up' from the Mallard, as the plumage is identical. They also eclipse moult once a year, where the drake temporarily adopts the duck's plumage. The Rouen, Duclair and 'natural' Mallard are the only breeds in which this occurs. Although the drakes' feathering may be the same as the ducks', drake bills always remain green.

Standards

The standards for plumage are the same as that for the Rouen.

Serious defects

A keel. Smaller than standard size.

Current status

We don't see many Duclairs at poultry shows, as most people seem to prefer the Rouen or the Appleyard.

Elizabeth

Origin

The Elizabeth is the only domestic breed of duck originating in Australia. It was developed by Lance Ruting in New South Wales and was named after his wife. It was bred as a small table bird for the Asian market. There is much controversy over the breed, its heritage and its status in the show arena. In some states they are not accepted for judging.

Figure 3.8: An Elizabeth duck.

Description

The Elizabeth is around the same size as a bantam Appleyard, but it is more compact and carries a lot more meat. They are brightly coloured, pretty birds. The drake's plumage resembles that of the Appleyard, while the duck is silvery-white with chestnut flecking.

The Elizabeth duck lays from 100 to 150 eggs a year. She is a fantastic mother and will go up your arm faster than a rash if you interfere with her offspring.

Elizabeths are not easily obtained, as breeders are few and far between. In fact, many people are not aware of their existence. I did breed Elizabeths for a short time, but they are flighty birds and can fly exceptionally well, so I found them very difficult to contain in one yard. The breed is quite a good forager and is economical to keep.

Standards

The drake's head is bottle-green, with the colour extending down the neck to a white ring that almost encircles the neck, but does not meet at the back. The eye is full, brown and bold. The back and rump are rich green-black, and the tail slate-black. The breast is deep claret, and the flanks, stern and abdomen French grey. The flanks are clearly marked with black, giving an appearance of chain mail. Across the wings are broad purple bands, outside of which are narrow bands of black and an outer band of white. The small coverts on the wings are rusty-red. The bill is slate and the webs are orange.

The duck has a ground colour of chestnut-brown peppered with black feathers. The head is a lighter chestnut that forms a perfect hood on the neck line. There is a purple-blue wing bar. The eye is dark brown, and the bill brown.

Serious defects

Green bill in drake, orange bill in duck.

Current status

Rare. This breed can be found, but not in the great numbers you would expect.

Indian Runner

Origin

The Indian Runner was introduced into Australia from Malaya more than a century ago. At the time, they were called Penguin Drakes, because of their long flat bodies and erect walk. The Fawn and White variety was standardised by the Indian Runner Duck Club in 1907.

Description

The outstanding feature of the Indian Runner is its unmistakable type and carriage; unlike other breeds, it has no pronounced shoulders and no great depth of body. In fact, viewed from the side, the body of the duck looks like an extension of the neck. When excited, the Indian Runner stands upright in a straight line from head to toe, giving it a lean and racy look. Indian Runners don't waddle, they walk.

Although it is no longer bred for egg production, the Indian Runner was considered *the* laying breed until the Campbell made an appearance. Individual egg-laying records were high.

Varieties of Indian Runner include the White, Fawn and White, Black and White, Black, Chocolate, Mallard, Trout and Blue.

Indian Runners are, in my opinion, one of the most intelligent breeds of duck. If you put the time into them, they bond easily with humans and make good pets. Mine know their names and respond when called.

Standards

Carriage should be upright and active. The angle of inclination of the body to the horizontal should vary from 50° to 80° when the bird is on the move and not alarmed, but when standing to attention, excited or trained for the show pen, it may assume a perpendicular pose. The body should be slim, elongated and well rounded, but slightly flattened across the shoulders. At the lower extremity, the frontline should sweep gradually round to the tail. The tail should be neat and compact and almost in a line with the body horizontally, but in some excellent birds is slightly elevated or tilted upwards. The neck should be long and slender. The head should be lean and racy looking, with a wedge-shaped bill. The skull should be flat on top, and the eye socket set so high that its upper margin seems almost to project above the line of the skull.

The eyes should be full, bright, and very alert and intelligent. The legs and feet should be set far back to allow for upright carriage. Plumage should be tight and hard. The total length of the drake should be 65–80 cm and the length of the duck 60–70 cm.

Serious defects

Above or below standard body weights and measurements. Body squat and short, oval or flattened. Domed skull. Plump head. Dished bill. Neck thick and short. Bulky trunk. Eyes set too low in the head.

Current status

Most colour varieties are plentiful. White varieties are the most commonly shown, while Blacks and Chocolates make an occasional appearance. Blue Indian Runners are rare.

Figure 3.9: Indian Runner.

Figure 3.10: Indian Runner, female.

Mallard

Origin

Wild Mallards are widely distributed, from eastern Asia to North America, and regularly visit almost every part of the world. Exactly how or when the domestication of these wild ducks took place is difficult to state, but it was definitely before the Christian era.

Description

This is the bird everyone tells us is responsible for some of our most impressive breeds. The name Mallard is thrown around so often, it is easy to forget that not everyone is familiar with this breed.

The Mallard drake has an iridescent green head, white neck collar, chestnut breast, silver underbody and blue wing bar. The duck's plumage is a combination of almond and golden browns with dark brown and black pencilling. She also sports a blue wing bar.

The great Italian naturalist Ulisse Aldrovandi called special attention to the striking peculiarity that Mallard drakes have four middle tail feathers curled

Figure 3.11: Mallard, male.

upwardly. Not surprisingly, with the exception of the Muscovy, all the drakes of the domesticated breeds mentioned in this book have curled tail feathers.

Mallards are available in both bantam and standard size. They are known as 'natural' Mallards, meaning they are true representatives of the wild colourings. (The Rouen has the same plumage.)

Obviously Mallards can fly, and the majority will already have been pinioned when purchased. The Mallard is shown Australia-wide and its colouring makes it very popular. They are prolific breeders, so it is not unusual for a duck to sit on 20 eggs and hatch all of them.

They do very well on dams. You will find if you purchase them as ducklings and place them on your dam, they will bond with that piece of your property and not fly away. A friend of mine has a restaurant and people can sit to enjoy their lunch and watch the Mallards he's raised swimming around on the dam below. They have become quite a drawcard for his business.

It is illegal to capture wild Mallards and/or gather their eggs without a permit. Licensing laws for keeping Mallards vary from state to state; for instance, it is possible to obtain a permit in Western Australia, but licences

Figure 3.12: Mallard, female and ducklings.

to keep Mallards will not be issued to private individuals in Victoria. Check with your state wildlife licensing organisation.

Standards

The drake's bill should be a bright yellow-green, with a black bean on the tip. The head and neck should be bottle-green down to a white ring that encircles the neck, but doesn't meet at the back. The eye should be bold, full and brown. The breast should be deep claret, while the flanks, stern and abdomen should be a French grey-blue. The flanks and sides should be clearly marked with black, like chain mail, as are the small coverts on the wings. The back and rump should be rich green-black, and the tail a slate-black. Across the wings are broad purple bands outside which are narrow bands of black then an outer band of white. The duck has a ground colour of chestnut-brown with each feather pencilled in black. The head should show a line from the beak to the neck. The duck's wing bar is the same as that of the drake.

Muscovy

Origin

The Muscovy is a native of South America. First European reports of the Muscovy didn't appear until the 16th century.

Description

There are still some doubts as to whether Muscovies should be classified as ducks or geese. Muscovies are certainly quite unlike any other breed of duck. For a start, they graze like geese. The males have no curled tail feathers; the feature that usually distinguishes sex in domesticated breeds of duck. Muscovies have no feathers around the face; instead the skin is bright red, and the drake has a knob on the head, which gives the appearance of a crest. Neither sex has a voice; the breed's sole means of communication is a hiss. To top it off, the incubation period of a Muscovy is 35 days, as opposed to 28 in others breeds, and if a Muscovy is mated to another breed the progeny are sterile. If I were classifying this breed I would not associate them with the duck family at all.

Not considered to be attractive by most, Muscovies are nevertheless hardy and quiet, and capable of foraging for most of their feed. They have been highly prized for their meat and make good children's pets, as they will bond with humans easily. They can be trained to know their names and will follow you around like a dog.

Although as a rule they possess a quiet disposition and make perfect pets, especially for young children, Muscovies will hiss at you and stamp their feet when you are out of favour with them.

Even if the appearance of the Muscovy does not particularly appeal to you, it is a valuable asset in your yard. In emergencies when your incubator runs out of room, use the Muscovy.

When sitting, the Muscovy duck can be very aggressive, and is not averse to applying the pressure of a strong bill around your arm, hand or wrist as she protects her young with vehemence. Not nervous, they make exceptional mothers and/or incubators. I have used Muscovies to hatch turkey, hen and duck eggs; all with excellent results.

Muscovy ducklings are very resilient and self-sufficient, preferring to do their own thing, often wandering away from the mother in their never-ending search for flies and insects. I have been entertained for hours watching them with their tails upright and their heads low and parallel to the ground as they give chase to insects.

Muscovy ducklings take less than 16 weeks to mature, and 7 months is the ideal age of a working drake. Because it is a heavy breed, after the age of 5 years drakes are not interested in riding the hens as much.

Muscovies are bred in White, Black and White, Blue, and Blue and White. Other colour variations are derived from these basic standards. The pattern of colour on each side of the bird should be a mirror image of the other. When purchasing a Muscovy, always look for good bone, broad shoulders, a rounded undercarriage and breast with no keel.

I have heard that some Muscovy drakes are sterile, or 'pelleted'. You will know if this is the case because the bird will have a 'baby face', and the red skin will not be as prolific around the bill and eyes.

Some of the lighter birds will fly, and fly exceptionally well. I had a duck that could circle 20 acres without any hesitation. She would perch on the roof of the house at night.

Unlike other breeds, they have a ritual for attacking the opposition. They wag their tails from side to side and sway their heads, hiss and laugh themselves into a frenzy, then those massive legs with long claws and strong powerful wings come slapping down upon the predator or opposition. Yet they can also be very sociable in little groups. It's as if one of them has told a brilliant joke and they are all highly amused as they laugh and bob their heads. It's contagious!

Depending on where you are in Australia, the size and weight ratios vary. In Western Australia the Muscovy duck has been inter-bred too many times, and the result has been a downsizing of the breed, much to its detriment. In eastern Australia the breed has shot ahead in leaps and bounds. The eastern bird dresses out to 5 kg, much more than the British poultry standards allow. In the west, we tend to refer to the eastern breed as 'mammoth' Muscovy, much to the eastern breeders' distaste. Personally, I prefer the mammoths.

Standards

The carriage should be low and jaunty. The body should be broad and deep, and very long and powerful. The breast should be full, well rounded and carried low. The wings should be strong and carried high, while the tail should be long and carried low to give a long appearance. The head should

Figure 3.13: Black and White Muscovy, male.

be large and adorned with a small crest of feathers, which are raised erect in excitement or alarm. There should be caruncles on the face and over the base of the bill. The bill should be wide, strong, of medium length and slightly curved. The eyes should be large, with a wild or fierce expression. The neck should be of medium length, strong and almost erect. The legs and feet should be strong, fairly short and set wide apart. The thighs should be short, strong and well fleshed. The feet should be straight, webbed and pronounced. The plumage should be close. The bird should feel hard, well fleshed and muscular.

Serious defects

Too small. A keel. Turned out wings.

Current status

Plentiful.

Orpington (Buff)

Origin

The Buff Orpington was bred by William Cooke of Orpington, England. The breed was developed by crossing a selection of Aylesburys, Rouens, Cayugas and Indian Runners. It was intended that the crossbreeding produce an all-purpose bird with good laying qualities, good table qualities and beauty.

Description

Orpingtons originally occurred in Blue, Silver, Black and Chocolate, but only the Buff is common today.

Buff Orpington ducks and drakes are both an attractive deep reddish-buff but, while the duck is buff all over, the drake has a seal-brown head. This colour finishes in a sharply defined line around the neck. The rump is red-brown. The bill in both sexes is orange and the legs bright orange. From a breeder's point of view, the Buff Orpington has the great advantage in that many of the best utility birds are also first-class exhibition birds.

They are excellent foragers and very vocal. The Buff Orpington is extremely placid. Two of our Buff Orpington ducks went broody and made a nest in an aluminium shed on which we were working. Both remained on their nests, in spite of the sparks flying everywhere, while I used an angle grinder to cut out a rather large window.

The Orpington lays a nice, well-shaped egg with green ones popping up from time to time in some strains. They lay an average of 240 white or tinted eggs a year.

Breeding the perfect duck or drake of this breed is not as easy as one would assume. The colour variations in the breed are unstable and blue feathers often appear. Some exhibitors feed keratin to their Orpingtons, which artificially gives them a deep buff colour. The downfall is that it makes their legs and webs a brilliant orange shade.

A blue or silver head in a drake is classified as a serious defect by poultry standards. But, if you happen to breed an Orpington drake with a blue or silver head, hang on to it. Although it may not be useful as a show bird, it will throw drakes with seal-brown heads, and good type ducks. It will also help to keep size in your birds.

There is a bantam version of the Orpington. Standards for the bantam version are the same as for the larger version.

Standards

Carriage should not be quite as horizontal as the Aylesbury, but neither should it be as upright as that of a Pekin or Indian Runner. The body should be long, broad and deep, without any sign of the keel. The back should be perfectly straight and the breast should be full and round. The wings should be strong and carried close to the sides, while the tail should be small and compact. The head should be fine and oval-shaped, with a narrow skull. The bill should be of moderate length. The eyes should be large and bold, and set high in the head. The neck should be of moderate length, slender and upright. The legs and feet should be strong and set well apart. Plumage should be tight and glossy, a rich, even shade of deep red-buff throughout, and free from lacing, barring or pencilling. The head and neck of the drake should be a seal-brown that terminates in a sharply defined line all the way around the neck. The rump should be red-brown in colour, as free as possible from blue. The duck should be similar to the drake's body in colour,

Figure 3.14: Buff Orpington.

and free from blue, brown or white feathers. The bill of both sexes should be orange with a dark bean. Legs and webs should also be orange. Eyes should have a brown iris and blue pencil.

Serious defects

Any colour other than stated. Twisted wings. Wry tail. Humped back or any other physical deformity. In the drake, silver-grey or blue head, white feathers on neck, brown secondaries, beetle-green on any part, very green bill. In the duck, very heavy lacing, strong line over eyes, white feathers on neck of breast, brown feathers, green bill.

Current status

Plentiful.

Pekin

Origin

Pekins originated in China, where they were bred as table birds. In 1873 a flock of Pekins was imported into Connecticut, USA, and the breed soon gained popularity. There is no evidence to show whether the breed travelled to England via the USA or China, but it is pretty clear that Australia obtained Pekins from England. The Australian bird has since been infused with strains of Indian Runner and White Campbell to improve its egg-laying qualities. This crossbreeding has lowered the body weight by overseas standards. The term Australian Pekin is used to differentiate between the original Pekin and the bird developed here.

Description

The Pekin is a large symmetrical duck, originally a creamy colour. It has a smooth outline, a characteristic upright carriage and broad round head. The bill should be rather broad and short for such a large bird. The skull should be broad and full, and at its round sweep in front should give the impression of a concave line from the top of the skull to the tip of the bill when viewed at a distance. However, do not get the view that concave is desirable, nothing could be further from the truth. The bill should be straight on top, or even slightly convex. The apparent concave appearance from the top of the skull to the end of the bill is caused by the general line from top to tip.

The eye should be rather deep set in appearance due to the broad skull, rather heavy cheeks and short, stout bill. This combination of features gives the characteristic Pekin expression.

When a mane appears on the back of the head and upper part of the neck of the Pekin drake, it increases the beauty and peculiarity of the male bird. This mane is seldom reproduced in Australia, but is highly prized when it does appear.

Supposedly, the Pekin is a nervous bird liable to heart attacks if chased. But chased senseless, the same thing would likely happen to us. A good Pekin is an absolute delight to observe and very attractive, with light yellow through its feathers. Despite their nervous disposition, if you move quietly around Pekins you have raised from ducklings, it is possible to bond with them.

My personal experience is that Pekins are very intelligent birds. They will walk into their house at dusk and wait for me to close the door. When someone enters their enclosure, they will stand behind the bamboo and observe what is going on. If you move from their view, they will come looking for you. They like to travel in little packs and will often gang up on anything unusual, such as a cat or dog walking on the other side of the fence. They quack loudly when things are not right in the yard, such as an invading crow, hawk or fox, and will chase lizards, especially blue tongues, which come in search of stray eggs.

If not harassed, Pekin ducks will sit and make good mothers, providing the environment is quiet and relaxed around them. Pekin ducklings are hardy birds with rapid growth. At 6–9 weeks they are termed 'green ducklings', meaning they are large enough to eat, but have not yet developed their pin-feathers.

The drakes like to know of any changes in the yard. They often follow me around observing my every move, and if they spot a shovel, that can mean only one thing ... worms! It's almost impossible to get rid of them if you are digging.

The Pekin is the premier table duck of North America and Australia. It produces an average 140 eggs a year.

Standards

Carriage should be almost upright, elevated in front and sloping downward to the rear. The body should be broad and of medium length, with no indication of the keel other than a little between the legs. The back should be broad. The breast should be broad and full, followed in line by the keel to a broad, deep paunch and stern carried just clear of the ground. Wings should be short and carried close to the sides. The tail should be well spread and carried high, the drakes with two or three curled tail feathers on top. The head should be large, broad and round, with heavy cheeks and a high skull that rises abruptly from the base of the bill. The bill should be short, broad and thick; slightly convex, not dished. The eyes should be partly shaded by heavy eyebrows and the bulky cheeks. The neck should be long and thick, and carried well forward in a graceful arch and with a slightly gulleted throat. Legs should be strong and stout, set well back and causing erect carriage. Plumage should be very abundant, and the thighs and fluff should be well furnished with long, soft, downy feathers. Colour should be cream or creamy-white.

Serious defects

Black marks or spots on the bill, although blemishes on the duck's bill are acceptable while she is in lay.

Current status

Plentiful in Australia.

Figure 3.15: Pekin, male.

Rouen

Origin

It is generally accepted that this breed originated in France, deriving its name from the city of Rouen. This part of Normandy has always been noted for the quality of its ducks and fowls. The suggestion that the original name of the breed was 'Roan' from the French word *rouan,* in reference to its colour, does not appear to carry much weight.

Description

The Rouen is, without a doubt, the most beautiful and imposing of the domesticated duck family. In colour and markings it is almost a duplicate of the Mallard, but it is larger and even brighter in colour than the Mallard.

Rouens are second only to Aylesburys and Muscovies in size. They are fair to good layers for a heavy breed. In body shape, the Rouen is like the Aylesbury; it has the same square shape, with the keel straight along the bottom, parallel to and almost touching the ground. The back is decidedly arched.

I have been successful incubating and hatching Rouen ducklings. The sooner you can get them in the great outdoors chasing mosquitoes and bugs the better. They double in size almost weekly, and before you know it, they are rather large and beautiful. The young drakes initially take on the colouring of the duck, but they soon colour up, and by 8–10 weeks are easy to pick by the colour of their bills, which should be green with a black bean, not blue. The duck's bill should be a dull orange with a saddle that extends from the head to midway down the bill.

They do look like little heavyweights, as the keel seems to be with them from the moment their bulbous bodies pop out of the eggs.

They are a curious duck, and so laid-back they are almost horizontal in their attitude. I have found that running these birds on thick grass improves their leg muscles and makes them more mobile. Rouens can be ornamental or exhibition ducks, although exhibition condition prevents them from successfully mating. They are not very fertile when they are carrying a lot of excess weight, nor are they free breeders, so you can expect no eggs, few eggs or unfertile eggs.

It is best to trim them down during the breeding season, the drake more so than the duck. They also prefer to mate on the water, as being such a massive breed, it can be difficult for the drake to tread the duck on dry land.

I have found that breeding a dark duck with the darkest drake results in good exhibition drakes. To breed good ducks, use a light-coloured and well-marked duck with a dark drake.

Rouens lay about 90 white to green eggs a year. Mine lay white eggs. If you wish to have only white eggs, keep breeding only from the white egg layers, or vice versa if you would rather have green eggs.

Rouens love to swim and move around; it is not true that a big duck doesn't need a lot of room. They are rarely still, except on really hot days, and then they can be found floating on the pond.

The adult Rouen drake moults into duck plumage in the summer. There is no bantam version of the Rouen.

Standards

Carriage should be horizontal, with the keel deep and parallel with the ground, and just clear of it. The body should be long, broad and deep. Wings should be large and well tucked into the sides. The tail should be very slightly elevated, and the drakes should have two or three curled tail feathers in the centre. The head should be massive. The bill should be long, wide and flat, and set in a straight line from the tip of the eye. Eyes should be bold. The neck should be long, tapering and erect; slightly curved but not arched. Legs should be of medium length, with stout shanks and should be well set to balance the body in a straight line.

Drake colouring

The drake's head and neck should be a rich iridescent green to within 2.5 cm of the shoulders, where the ring appears. The ring should be perfectly white and cleanly cut, dividing the neck and breast colours, but not quite encircling the neck, leaving a small space at the back. The breast should be rich claret in colour, coming well under, cleanly cut and not running into the body colour. It should be free from white pencilling or chain armour. The flank should feature chain armour, or flank pencilling, in rich French grey-blue, well pencilled across the glossy black, free of white, rust or iron. The stern should be the same as the flank and very boldly pencilled close to the vent, finishing in an indistinct curved line, followed by rich black feathers up to the tail coverts. Tail coverts should be black or slate-black with a brown tinge. The back and rump should be a rich green-black from between the shoulders to the rump. Large wing coverts should be a pale clear grey; small wing coverts should be French grey, finely pencilled. Pinion coverts should be dark grey or slate-black and should feature two bars. One bar should be composed of one line of white in the centre of the small coverts. The other should be a grey-tipped-with-black line at the base of the flight coverts. The latter feathers should be slate-black on the upper side of the quill and rich iridescent blue on the other side. Each of these feathers should be tipped with white on the end of the lower side, forming two distinct bars. Markings throughout the whole bird should be cleanly cut and well defined. The bill should be bright green-yellow, with a black bean at the tip. Eyes should be dark hazel. The legs and feet should be dull orange.

Figure 3.16: Rouen, male.

Duck colouring

The duck's head should be a rich, golden chestnut-brown with a wide brown-black line running from the base of the bill to the neck, and very bold black lines across the head, above and below the eyes. The two bold black lines should be filled in with smaller lines. The neck should be the same colour as the head, with a wide brown line at the back from the shoulders, shading to black at the head. There should be two distinct white wing bars with a bold blue ribbon mark between. As in the drake, the flights should be slate-black with a brown tinge, and no white. The remaining plumage should be golden chestnut-brown, with every feather from the throat and breast to flank and stern distinctively pencilled, the markings rich black or very dark brown, the black pencilling on the rump having a green lustre. The bill should be orange, with a black bean at the tip and a black saddle extending almost to each side and about two-thirds of the way down towards the tip. The eyes should be dark hazel, and the legs and webs dull orange.

Serious defects

Leaden bill. No wing bars. White feathers. Stern broken down. Wings down or twisted. In the drake, a black saddle, black bill or minus a ring. In the duck, a white or approaching white ring on the neck.

Current status

Plentiful.

Saxony

Origin

The Saxony duck existed in 1934 in Germany. It is believed to have been bred from the Blue Swedish as a dual-purpose duck.

Description

The Saxony drake is coloured like a pale Mallard. It has a pigeon-blue head and neck, and a white ring around the throat. The lower part of the neck, shoulders and breast are a rusty-red, with a hint of silver lacing on the breast. The back and rump are dark blue, as are the wing bars. The tail feathers are fawn or cream. The duck is a soft buff colour, with creamy wings and light blue wing bars.

The plumage of both sexes has a grayish-blue cast, and in both sexes the bills are yellow, with a green shading in the drakes and brown in the ducks. Eyes are dark brown, and legs and webs dark yellow.

There is a bantam version of the Saxony. Standards are the same for it as they are for the larger version.

Standards

The Saxony has a strong carriage, with a long broad body and absolutely no keel. The body should be long and stocky, with a broad back sloping slightly to the rear. Wings are short and held close to the sides. The tail is long, carried full and closed. The head is long and flat, and the neck is solid. The bill is of average length, but broad. Legs should be set in the middle of the body. The plumage should be tight, with soft underfeathers. The drake's head and neck should be very blue, with a white ring around the throat. The lower part of

Figure 3.17: Saxony, female.

the neck, shoulders and breast should be rusty-red, with a hint of silver lacing on the breast. The back and rump are dark blue. Wing bars are purple-blue. The tail feathers are fawn or cream. The duck's head, neck and breast should be buff in colour, with a white line travelling through the eye and a white ring on the neck that doesn't meet at the back. The back and breast should be a light buff; wing bars and tail, a light blue. Wings are cream in colour. In both sexes the bills are yellow, with a green shading in the drake's and brown in the duck's. Eyes dark brown, and legs and webs dark yellow.

Serious defects

Poor colouring, wrong wing bar colouring. Wrong bill colour. Drake's head the wrong colour, ring on neck joins at back. Duck too dark in colouring, resembling an Appleyard.

Current status

Stable.

Welsh Harlequin

Origin

The Welsh Harlequin is a colour sport from the Khaki Campbell.
Again, I don't believe this should be termed a 'purebred'.

Description

The Welsh Harlequin duck has a dark slate-green bill and a green wing bar; part of its Khaki Campbell inheritance.

The same characteristics should be noticeable in the Welsh Harlequin as in the Campbell: a long body, fine head and slender neck. They are active, good foragers, and can fly, although not as well as the bantam Appleyard. The drake's bill is greenish-blue, and the duck's greenish-slate. The drake's legs are orange and the duck's orange-brown. Both sexes have green wing bars.

A good strain of Harlequin can equal a Campbell in egg production, and has the added advantage of being an excellent table bird. It is docile and placid, with less tendency to flap and panic than many other breeds, and has attractive plumage.

Standards

The drake's head and neck have black-green plumage with a bronze lustre, and a white ring encircles the neck. Claret feathers with green flecks extend down the breast and shoulders. The keel is cream. The tail and rump are the same colour as the head. The duck has a fawnish head with brown flecks, and a ring around the neck. The breast is flecked in red-brown claret, and the undercarriage is white. There is a green wing bar on the brown-red and black of the wings. There are green flecks through to the tail.

Serious defects

Oversize, wrong bill colour, wrong wing bar colour.

Current status

There are plenty of Welsh Harlequins around, but good Welsh Harlequins are rare.

Figure 3.18: Welsh Harlequin, male.

4
Selecting a duck

Selecting a breed of duck is probably the most difficult thing to do. There are as many attractive breeds of ducks in Australia to choose from as there are practical ones. Ultimately your decision should be guided by why you want to keep ducks. Do you want to keep them for eggs? For meat? As pets? Just because they're beautiful? To show? To breed? To keep the garden free of slugs and snails? Or for a number of these purposes?

Some ducks are bred primarily for egg production, others for meat, and some are particularly ornamental. Some breeds are nervous or flighty, others are placid. Some ducks make good mothers, others are non-broody.

Other factors will also affect your choice. If space is at a premium, obviously a smaller breed of duck will be best. Elizabeth ducks are small, as are Black East Indians. There are also bantam versions of some of the larger ducks, the Silver Appleyard, for instance. But, be warned, these smaller ducks can fly extremely well – like over the hill and far, far away. So, if you intend running these breeds, ensure that they are pinioned or that the primary flight feathers on one wing are cut short. This will keep them on the ground and in your yard.

Bantam ducks always remind me of little garden ornaments, they are cute and colourful, and often afraid of very little. They do make good pets when raised from ducklings, and really a duckling is the best way to buy a bantam duck, as it will grow up viewing its surroundings as home and be less inclined to fly away as an adult. This is an advantage because you probably won't have to cut the wing feathers, so the duck will still be able to fly from danger if a stray dog, cat or fox confronts it.

Egg-laying ducks

One of the best reasons for keeping ducks is that they can supply you with eggs.

Properties of a duck egg

Eggshell colour: can vary from white to green in some breeds, and even to black.

Fat content: 14%, compared with 11.6% for chicken eggs.

Weight: from 50 to 80 g.

Yolk: ducks that are fed grains and meals (e.g. fishmeal, soya bean meal, meatmeal) will lay a pale yellow yolk. More animal protein in the diet produces a lemon to yellow yolk. Free-ranging ducks lay eggs with a deep yellow yolk.

Taste: hardly varies between duck and chicken eggs, unless the ducks have been ranging free and eating snails, frogs and bugs. Then their eggs have a strong gamey flavour.

Cooking: duck eggs contain more fat, which makes them perfect for making cakes. They fry and boil more quickly than chicken eggs. When whipped, the white of a duck egg does not become fluffy like that of a chicken egg, nor does it remain stiff, so you can't make pavlovas out of duck eggs.

Nutritional information: duck eggs are higher in cholesterol, fat and calories than chicken eggs.

Table 1. Composition of eggs of various poultry species

Bird	Protein (%)	Fat (%)	Cholesterol (%)	Calories per 100 g
Chicken	12.9	11.5	4.94	163
Duck	13.3	14.5	8.84	191
Goose	13.9	13.3	–	185
Turkey	13.1	11.8	9.33	170

Egg-producing breeds

Indian Runners and Welsh Harlequins are all light egg-producing breeds. Khaki Campbells, in particular, can be terrific egg producers; a good layer laying up to 360 eggs a year. However, if you want a good egg yield, make sure you are buying an authentic Khaki of a good egg-laying strain. So much emphasis has been placed on showing the breed that some strains have lost some of their egg-laying abilities. In addition, a great number of Khakis on the market today have been crossed with other breeds, resulting in the deterioration of egg production. Khaki Campbells are also good broodies.

Indian Runners were the best egg layers before the Khaki Campbells were developed. They also make good broodies.

If you're interested in coloured eggs, some breeds lay eggs with quite distinctive shell colours. For example, the Pekin ducks that I have lay a white egg with a subtle tint of pink. The Rouens lay an egg with a whiter appearance. The Black Indian Runners lay a grey egg, sometimes with black dots on the shell. The Muscovy egg has a greenish tint and the Aylesbury egg could easily be mistaken for a goose egg in size. Some breeds of ducks will lay a green egg. It is my opinion that these ducks have had a Mallard introduced into their line, as Mallards lay a green egg, mostly for camouflage in the wild. The Khaki Campbell is a prime example of this, and Mallards were used in developing this breed.

Meat ducks

Duck meat is tasty and lean, and these days many people want to keep ducks for meat. What they need is a breed that has been developed as a meat producer, particularly one that grows fast. Remember, the faster they grow, the less food they eat.

Meat breeds

Aylesburys, Duclairs, Elizabeths, Pekins and Rouens were all developed for the table. Domesticated Mallards are also classed as meat birds.

Alyesburys, Duclairs, Pekins and Rouens are all classed as heavy birds; the adult drakes weighing in at 4–5 kg. Mallards and Elizabeth ducks are much smaller, weighing less than half that of the larger meat birds. In fact, the Elizabeth duck was developed as a small table bird for the Asian market. The Aylesbury dominated the table duck trade in Britain for many years. It is a large, placid white duck that lays an egg the size of a young goose's egg. For a heavy breed, the Aylesbury is not a bad layer either, laying up to 150 eggs a year.

The Pekin is the premier table duck of North America and Australia. Pekins and Aylesburys are the only ducks that produce white meat. Pekins are also the fastest growing of all the breeds. They are not good broodies and only average layers, producing about 140 eggs a year.

Even though in Australia Rouens are predominantly an exhibition bird, they were originally bred as a table duck. The Rouen combines stately beauty and grace. They are slow growing and only average layers.

Dual-purpose breeds

Dual-purpose, or utility, ducks are those that produce both meat and eggs. There are a number of dual-purpose breeds, for it makes sense to keep ducks for more than one purpose. Appleyards, Blue Swedish, Cayugas, Crested Ducks, Muscovies, Orpingtons and Saxonys are all dual-purpose breeds.

The black-plumaged Cayuga is a general-purpose duck. As well as being good meat birds, they are outstanding layers, but not fast growers. Crested Ducks, on the other hand, are fast growers, but have a low hatch rate. The Buff Orpington is a very beautiful dual-purpose duck, and the deep buff plumage is quite stunning. They are also very good broodies.

To my mind, the Muscovy is one of the best dual-purpose ducks around, although it is also one of the most misunderstood breeds. This bird makes an ideal child's pet, is quiet, a good layer, a brilliant incubator and a great meat bird. Many view them as ugly because of the caruncles (red skin) around their face.

Figure 4.1: A pair of Muscovies, owned by Nyiri Murtagh.

Table 2. Duck breeds: purpose, weight, production, temperament and mothering ability

Purpose	Breed	Class	Drake weight (kg)	Duck weight (kg)	Eggs per year	Temp-erament	Mothering ability/ broodiness
Egg production	Campbell (Khaki)	Light	2.25–2.5	2–2.5	250–330	Placid	Fair
	Indian Runner	Light	1.6	1.4	225–330	Nervous	Excellent
	Welsh Harlequin	Light	2–2.5	2–2.3	240–320	Placid	Good
Dual-purpose or utility breeds	Appleyard (Silver)	Heavy	3.5–4.9	3.2–3.7	25–125	Placid	Excellent
	Blue Swedish	Heavy	3.7	3.2	150–180	Placid	Good
	Cayuga	Heavy	3.6	3.2	150–230	Placid	Good
	Crested Duck	Light	3.2	2.7	100–175	Placid	Fair
	Muscovy	Heavy	4.6–6.4	2.3–3.2	200–270	Placid	Excellent
	Orpington (Buff)	Heavy	2.25–3.4	2.25–3.2	200–300	Placid	Excellent
	Saxony	Heavy	4.1	3.6	150–170	Placid	Fair
Meat	Aylesbury	Heavy	4.5	4.1	35–150	Placid	Poor
	Duclair	Heavy	4	3–3.5	170–200	Placid	Good
	Elizabeth	Bantam	2	1.5	100–150	Flighty	Good
	Pekin	Heavy	4.1	3.6	160–230	Nervous	Fair
	Rouen	Heavy	4.55	4.1	35–150	Placid	Good
	Mallard	Small	1	0.7	160–230	Flighty	Excellent
Ornamental	Black East Indian	Bantam	1	0.6–0.9	25–125	Flighty	Excellent

The information above is based on my experience with these breeds. The figures are estimates and may vary according to the conditions under which the ducks are kept, including their diet.

5
Buying a duck

When you have decided you want a duck, regardless of whether you intend joining a poultry club to show the bird, or simply to have it running around in your backyard, there are a few things you should look for.

It is always best to buy at least two; that is, a pair or two ducks. If you are purchasing ducks for breeding, the best results are gained with trios comprising two ducks with one drake. In my experience, the more ducks you have, the fewer eggs you get. Trios work well; you will always get two eggs from two ducks.

Be sure of the breed you are buying, do your homework and know what the duck and drake should look like. Don't always rely on the person selling the birds. In buying a breed that I was unsure of, I have been duped and ended up with something different from what I paid for.

I made my first big mistake with a pair of Buff Orpingtons. I had not done my homework and did not know the breed. I bought from a 'reputable' breeder, only to discover they had been crossed with Khaki Campbells, hence the bills and legs were black, and the feathers a pale, dirty shade.

Here are some questions I suggest you find answers to before buying ducks.

Where did the breeder obtain the original stock?

Was it from a breeder or a home for lost ducks? This will give you some idea of the bloodlines behind the bird.

How old are the birds?

Are they ducklings? If so, how do they look? Healthy and vibrant, or slow and lethargic? If they appear unwell, don't purchase them. The ducklings may die when you take them home.

What should you look out for?

If you are buying a mature bird to lay eggs, here are some useful tips for choosing a quality, healthy bird that will meet your requirements.

(1) Check the vent. There are two laying bones either side of the vent. They are quite pointy and should be at least 7.5 cm apart. This means that the bird is in full lay.

(2) Check the duck's bill colour. Is her bill a lighter shade than those of the other ducks? This also points to the fact that she is laying. Don't forget, it is not uncommon for a laying duck to have blemishes on her bill.

(3) You will know if the duck has been ridden by a drake because there will be feathers missing down the back of her neck. Check to see that she isn't red and bloody on her neck from over-riding. Sometimes this can cause prolapse in the duck.

(4) Look under the wings and brush the feathers back to see if there are any lice or mites on the bird. Look, also, for small, black seed ticks, because these can cause paralysis in the bird.

(5) Check around the eyes for signs of weepiness and stickfast flea.

(6) Look at the undersides of the feet. The feet should be smooth with no calluses, sores or anything else on the underside. Are there any corns? If the duck has corns, it can mean one of two things:

    (a) The bird has been running on hard ground and is approximately 2 years of age. (Some breeders believe that corns on the feet are a good indication of a duck's age as a bird running on hard ground will take 2 years to develop corns that bleed. Personally, I disagree with this theory. A duck reaches maturity at the age of two and can live for 10 to 15 years, so why would you want a duck in its prime with sore feet?)

(b) The duck is suffering from some sort of deficiency, usually niacin or riboflavin, and will not make a good breeder. According to Dave Holderread, in *Raising the Home Duck Flock*, 'The most common cause [of corns], although not usually recognised, is a dietary deficiency in biotin, pantothenic acid or one of the other vitamins or minerals that are important in maintaining healthy skin'.

(7) Look at the duck's nostrils to ensure they are free from any obstruction. They should be dry and clean. Be sure the duck's vent is clean and not soiled.

(8) You can tell a lot about a duck by the feel of it. It should be soft to touch. Young birds have a marshmallow-like feel to them.

Ageing a duck is very difficult. Unless you know the breed exceptionally well, you are at the mercy of the breeder. However, once you get the duck home and it begins to lay, its age should shine through in the laying pattern. If, for example, you have bought a 4-year-old duck, she will probably lay an egg every 4 days. If the duck is 6 years old, then she'll lay an egg every 6 days. If she is a young bird, say 1 or 2 years old, you should receive an egg every day. Hybrid ducks (crossbreds) will lay more regularly.

Having said that ageing a duck is difficult, I have found that careful study of the rim and corner of the eye can give an approximate indication of age. A young bird will have a very bright full eye, and the lid of the eye will be taut. As the bird grows older, the lid or rim of the eye grows loose and the white from the corner of the eye moves in, pushing the exposed eye back a little.

Mature ducks (2–3 years old) are excellent producers of fertile eggs, they are good mothers and magnificent to observe. I support the theory that it is worth controlling the diets of the heavier breeds, such as the Rouen and the Aylesbury, so they do not become overweight, as this inhibits their breeding ability. Make it a practice not to buy a duck older than 2 years.

A healthy duck will be bright and will try to get away from people. This is especially true when it arrives in its new environment. The plumage should be smooth and clean, especially in the vent area. A new duck may be dirty because it soiled itself on the way home, which of course should not be interpreted as a sign of illness.

The state of nutrition can be determined by feeling the bird's breast region. The breast muscle should be well developed. Ducks that are sick or not properly nourished have a narrow breastbone that protrudes sharply. However, with baby ducklings you cannot determine this, as their muscles have not yet developed.

Importing waterfowl

In the 1940s, it was possible to import ducks into Australia. This is something for which we can count ourselves lucky, as it has meant we have many breeds to choose from today.

Importing waterfowl into Australia is now illegal, and I doubt the laws restricting this practice will ever change. There are many reasons for this, the main one being that waterfowl are possible carriers of influenza viruses that can be contracted by humans.

You can apply to the Australian Quarantine and Inspection Service (AQIS) for an Import Query Result regarding the importation of fertile domestic duck eggs. The sheer number of forms and statutary declarations that arrive on your doorstep will amaze you.

The documentation/process involves:

- Obtaining prior permission, in writing, from the Director of Animal and Plant Quarantine to import fertile duck eggs.
- Submitting written applications through the Chief Quarantine Officer of the state in which the eggs are to be quarantined.
- Consigning eggs to the address of the Australian importer, care of the Chief Quarantine Officer of the state in which they are to be quarantined. All consignments must be accompanied by a Permit to Import and the appropriate Zoo Sanitary Certificates, which must not be modified without written permission of the Director. The Permit to Import, together with the documents listed, must be produced to a quarantine officer at the port of entry before an Animal Quarantine Entry will be issued.
- Ensuring declarations are filled out by a government veterinary officer of the country of export.
- Ensuring eligibility for you to be considered for importing is taken into account and examined.

- Quarantining the source flock (the flock of ducks overseas from which you intend to gain your eggs) for 90 days prior to collection of eggs.
- Housing the source flock in a rodent-free environment.
- Collecting, packing and transporting eggs under the supervision of a government veterinarian. The package must not leak and must still be sealed when it arrives in Australia.
- A first certificate from a government-approved veterinarian of the country of export.
- A second certificate from a government-approved veterinarian of the country of export.
- A certificate from the veterinary officer in charge of the designated government quarantine facility.
- An Aviary Importation Program, which approves tests for the importation of fertile eggs of fancy breeds (of duck).
- Comparison of AAHL (Australian Animal Health Laboratory), SCAHLS (Subcommittee on Animal Health Laboratory Standards, Australia) and OIE (Office International Des Epizooties) protocols and tests.
- Interpretations.
- Hatching egg importation program.
- Hatching sampling for bacteriology culture of *Salmonella* spp.
- Tests on dead-in-shell embryos.

As an importer you should be aware that:

- Should the container be damaged, show signs of leakage, or have inadequate certification, the consignment will be refused entry into Australia.
- The eggs will be hatched at your expense in isolators in a government quarantine facility designated by the Director. The hatched ducklings will remain in quarantine for 12 weeks.
- The shells and membrane and ducklings will undergo extensive pathology tests at your expense. Unhatched ducklings will also be tested for disease. Action will be taken should any of these pathology tests prove positive. These actions may date back to the original source flock, with a vaccination history.

Add to all of the above the costs of the incubation, the room you use, costs of quarantining the source flock and so on. This is a very brief summary of your costs and responsibilities. In short, importing duck eggs is a very expensive and unproductive exercise. And at any time your consignment of eggs may be refused entry or terminated ... and you still incur all the costs.

6
Duck behaviour

Foraging

The wonderful thing about ducks is that they don't dig up your garden. And yes, all ducks eat snails, not just Khaki Campbells. I think that because Khakis were once very popular in Australia (there was a time when at least one house in every street owned a couple of Khakis that kept mum's garden free of snails), most people grew up with the belief that Khaki Campbells were the only snailers. Not true. Ducks are foragers by nature, and anything that makes the mistake of crawling or flying past a duck is viewed by that duck as lunch!

Ducks can see 360 degrees without moving their heads. This works to their advantage when there is eminent danger, or in spotting something yummy flying past. They don't dig holes in your vegetable patch; however, they are partial to young green shoots.

Ducks naturally sift through mud, foraging for tiny food particles with their bills. The bill is perfectly designed for this. The mandible is the upper broad and flat part of the bill. Inside it is covered with soft skin and tiny taste buds.

The nerves leading to the taste buds run through tiny pores in the bone of the mandible. Because of this, it is extremely sensitive and it makes it possible for ducks to sense and catch tiny organisms in mud.

Along the outer and inner edge of the lower mandible there are fine pointed plates, called 'lamellae', resembling little serrated teeth. The tongue has similar serrated edges. When a duck picks up mud, it closes its bill and

Figure 6.1: A pair of Buff Orpingtons, a breed of excellent foragers.

uses its tongue to force the water and mud out between these lamellae. Solid matter is retained.

Of course, ducks feed on land as well as in water, breaking off succulent green shoots, chasing insects, frogs, grasshoppers and flies, and dining on any fallen fruit. They love mulberries! I often watch my ducks on the pond in the afternoons as they jump into the air catching small flying insects with great zest.

Intelligence

Ducks are intelligent and it doesn't take long for them to realise where the food is kept and who brings it to them. Soon they will begin to quack loudly in anticipation whenever they see you heading towards the shed where you store the grain.

If given the opportunity, they will help themselves in a moment's notice to the stored supplies. I have seen ducks continually rip away at the bottom of grain bags until they tear a hole in them, then sit and eat the grain as it spills out in front of them.

A friend of mine had Muscovies that watched his dog place its paw in the backdoor and pull it open. Within minutes the Muscovies opened the door with their bills and raided the pantry, much to my friend's horror and amazement.

Indian Runners are known for their intelligence and they don't forget anything. Ebony, our Black Indian Runner drake, worked out how to run up the fence that partitioned him from the Pekins on the other side. He would take a run at the fence, scale it and do a little more than 'talk' to the 'chicky babes' on the other side. When he heard our footsteps approaching, he would make haste back to his own side of the fence and jump in his pond. We had absolutely no idea what had been going on, until we began hatching Black Pekins!

Needless to say, we moved Ebony to a more secure yard. Now he spends his time calling out to any drakes that pass by. When they make the fatal mistake of getting too close to the wire he carefully pulls out all their chest feathers. The silly drakes come away bald-chested and ugly.

But it's just not the above-mentioned breeds that have a story to tell. I have found that all breeds of ducks can give you hours of entertainment, moments of tremendous humour and endless tranquility.

Flight

Many of the bantam breeds 'fly like a bird'. Even some of the Muscovy ducks fly exceptionally well. I have experienced Muscovies flying off and disappearing over the horizon. I often joked that if they flew over our house one more time I would get out the shotgun. Some breeders will tell you that Indian Runners fly. I have never seen our Indian Runners take to the air. The Rouens have no chance, nor do the Pekins; dropped from a roof, they would fall like a rock. Their little wings could never support their big bodies in the wild blue yonder. In fact most of the standard breeds of ducks do not fly, just as a decent Muscovy should not be able to fly. (Standard breeds meaning large breeds, as opposed to bantam breeds.) The tail feathers act as a brake and as a rudder for steering. In addition to forward flying, ducks soar (glide on wind currents) and hover.

Most small ducks take off with a quick upward leap into the wind and strong, fast wing beats. Dabbling, surface-feeding ducks also jump directly from the water. Many waterbirds, however, have to run over the surface of the water with wings flapping until they gain enough speed to lift themselves into the air. Birds land by setting their wings at the proper angle to reduce speed and then throwing the body backward and the legs forward.

Preening

There is an oil gland, called the uropygial gland, at the base of a duck's back, where the tail feathers begin. The ducks squeeze this gland, stimulating the release of oil and then place it on the feathers by individually running them through their bill. This oil assists in the feathers shedding water.

In some breeds, the Muscovy being one of them, there's a bright yellow patch around the base of the tail, especially during the mating season. In the height of the breeding season, the drakes spend a good deal of time preening, exciting the oil gland and leaving a yellow mark on the feathers.

Ducks and water

Ducks are waterbirds, and water plays a very important part in their lives. From the day they pop out of the shell, they love to play in water, so it is important for your ducks to have a pond or bathtub in which to swim and preen. Water also encourages them to mate. But you must ensure that they are easily able to climb out of the swimming hole. If they can't get out, they will become exhausted and sink. Ducks can drown!

Mating behaviour

Ducks often become very excited and playful with one another when their pond is being filled. Observing your ducks in their swimming pond, you will notice that drakes generally become active first thing in the morning and late in the afternoons. They have a little ritual dance, where they bob their heads up and down as though they were at a disco. This dance seems contagious, as all the drakes participate. They then swim as they wait for the others to join them, occasionally stretching upright in the water while making a drumming noise.

Figure 6.2: A pond is a great aid for fertility.

When the ducks enter the water, they dip their heads, each bird splashing the water over herself and spreading her tail feathers. Often her tail will be erect and pointing skyward. Lifting both wings into the air, she then stretches her primaries. This performance goes on for some minutes. Then, bobbing her head up and down, she swims closer to the drake, then swims in tight circles around him. Next, wild dashing about on the water begins as they both dive and skim across the water.

Occasionally the birds will exit the pond in a mad frenzy, wings outstretched and flapping wildly, returning post-haste to the water. Initially, it may appear that they are becoming aggressive, but they are bathing in preparation to mate and enjoying every second. I view this as natural mating when the drakes become very amorous towards the ducks.

Some drakes are a little more in need, and will chase the ducks as fast as their webbed feet can carry them across dry land. Sometimes this incites the other drakes so they too want to be in on the action.

Catching the duck, the drake mounts her and grasps the feathers directly behind her head. Sometimes the drake appears to be drowning the duck, submerging her head for what seems like a long time. He flaps his wings sometimes for balance, then brings his vent (cloaca) into direct contact with the duck's. He then inserts his penis, or pencil, into the cloaca of the duck.

Muscovy drakes often laugh and wag their tails at the ducks when they are about to mount. There is little second-guessing as to what they are up to. Ducks will pair off, and it is not unusual for drakes to have favourites.

I bought my first duck some 10 years ago. He was a Muscovy drake named Donald Duck. He had been raised by a hen and when I brought him home, he suffered a serious identity crisis. Donald believed he was a chicken and that the ideal girlfriend should have a wattle and comb, hate water and be white.

I bought a Black and White Muscovy duck, named her Daisy, of course, and introduced her to Donald. Donald couldn't care less about Daisy. After some lengthy discussions I decided to obtain a white duck. With fingers crossed I placed her delicately in the yard with Donald. I honestly can't say it was love at first sight, but he did give her more than a glance and we were on the way to a very successful union and the plodding sounds of little webbed feet.

Nesting behaviour

Most ducks can only cover and turn a clutch of about 14 eggs, though the Muscovy, being bigger, is capable of hatching considerably more. I often use my Muscovies to hatch turkey and goose eggs, as they are brilliant incubators and mothers.

Typically a duck will lay and sit on a clutch of 12–20 eggs, providing you don't keep collecting them. Some ducks lay excessive numbers of eggs. I know of an Appleyard that laid more than 40 eggs and attempted to sit on them. Too many eggs in a clutch will result in a very low hatch rate, so remove most of these eggs from the nest.

A tyre is an excellent place for a duck to lay. Simply line it with straw. This will prevent other birds scratching around in her nest and will keep the ducklings close to her if they hatch at night. Sometimes ducklings hatch and decide to go for a walk on their own, especially Muscovies.

Figure 6.3: A sitting Muscovy.

It is very easy to spot a broody duck, she will usually be the noisiest and her feathers will look as though they are standing up. If your birds range free, start looking for a nest.

It is best not to disturb a duck while she is sitting. She will seem very unsociable and may become aggressive when approached. (Ducks stand their feathers on end and quack defensively if they feel threatened.) Continual disturbances may cause her to abandon her nest and the eggs.

When your duck sits, make sure she is safe from foxes, dogs and crows. If she commences to sit outdoors, move her so she doesn't fall prey to predators, though this may not be easy.

A duck leaves her nest every couple of days for some food and a swim. When she returns, she stands over her eggs, sprinkling them as she preens the excess water from her feathers.

Don't be surprised if a duck rolls an egg out of her nest and away from her. They know if their eggs are infertile or damaged and will often push them out.

When I first began running ducks, I noticed an egg had rolled out from under one of my ducks, I laughed at her and said, 'You silly thing', as I lifted her gently and placed the egg back under her. The following day, the same thing happened. Once again I placed the egg under the placid duck. Finally the egg exploded under her. One thing was for sure, the look she gave me left no doubt as to which one of us was stupid.

During the final days before her ducklings hatch, she will be reluctant to leave the nest. At this stage, in summer, it is a good idea to place some food and water beside her. Then the greatest miracle of all: little webbed feet, and cute and curious faces will appear from beneath and between her feathers.

For the first couple of days the duck will remain on the nest. This is to allow time for the ducklings to dry out and find their feet.

Ducks are placid creatures with a gentle disposition. The only time a duck will grow slightly aggressive is when she is sitting on her eggs or has ducklings, and this is an act of self-preservation and protection for her young.

Often a duck won't take her young ones out of the shed for a while. She knows that outside there are crows and other predators that will pounce on an unsuspecting duckling.

Make sure that the ducklings' feet can touch the bottom of any dish of water you provide for them. They will certainly try to swim in it, so a shallow dish will ensure they don't drown.

Moulting

Moulting, or the replacement of old feathers with new ones, normally takes place at the beginning of summer and may last for 6–8 weeks. Egg production will drop, and may even cease during this period.

When the duck moults, the laying bones draw close together and you will know she isn't passing eggs. During the moult, ducks will be using all their nutrition to grow new feathers. It is not uncommon for them to go broody and hatch eggs at this time of the year. The duck will use loose feathers to line her nest.

A drake will lose his brilliant plumage and the curled feathers at the base of his tail will fall out. In some breeds (the Rouen, Mallard and Duclair) the drakes will 'eclipse moult', meaning they will colour up the same as the female, then grow their brightly coloured and curled feathers just in time for the next breeding season.

Changing sex

This is a fallacy. Ducks do not change sex. I have heard of cases in which a duck will colour up like a drake, and will even acquire tail feathers and cease to lay, but believe me, she will not service other ducks. There could be many reasons for a duck to undergo these changes. It could be that this is a lonely duck with no company that doesn't feel the need to fulfil her lot in life and lay. Who knows? But the fact remains, she is not a drake.

7
Duck management

Flock management

If your ducks run free, when you first get them home it is important to let them adjust to their new environment. Keep them locked up in their house for 2 or 3 days to give them time to know where their food and water is, followed by a few weeks penned in their yard during the day until they settle in. You will then find they can be given free range during the day and they will return to the safety of their housing each night of their own accord.

Some breeders choose not to let their birds out until after they have laid: ducks normally lay their eggs before 9 a.m. However, I like to see my birds out at sunrise and find that the ducks will return to their houses in the course of the morning to leave me their precious eggs.

You shouldn't have any trouble getting your birds back into their enclosure. Ducks are creatures of habit. Once they know where they live, they will put themselves to bed at dusk. It will simply be a matter of walking over and closing the door.

When breeding season is over, all the birds should run together where the grass is green and long. It's what I call R & R time for the ducks.

I keep the older birds accustomed to seeing young ducklings, and the young of other fowl, by placing young ducklings (5 weeks and older) among them. I run my geese, hens and ducks together on green grass all year round, and find they live in harmony.

Figure 7.1: Ducks, fowl and geese can all be run together, except during the breeding season.

Ducks can be very territorial and unaccepting of new or introduced birds, but if they grow accustomed to seeing ducklings running about, they are more even-tempered and accepting when you introduce a new breeder into the flock. Otherwise it can be a nightmare.

Breeding management

Purebred birds should be separated from other breeds 3 weeks before the breeding season.

Quite a few breeders speak of 'flock mating'. This refers to the practice of running a surplus of drakes with the ducks, say one drake to every duck. I haven't found this to be necessary, although I do run pairs of Rouens, simply because they have chosen to pair up and bond with one another.

Flock mating can be risky, especially if all the drakes have a favourite duck. This duck will be ridden by the drakes one after the other, which will eventually result in a prolapse, where the duck's reproductive organ protrudes through her vent. This can be fatal (see 'Remedies and cures').

A duck may also be ridden by a drake to the point where she is bloody around the back of her head. If you notice this happening, it is advisable to remove the duck, placing her in another yard for recovery. Terramycin can be sprayed on her wounds to promote rapid healing.

Over-riding can also result in bruised muscles in the duck's legs. An affected duck will be bright enough, but either unable to walk or suffering a serious limp. Remove her from the yard and rest her until she has recovered. Her eggs will be fertile for a minimum of 10 days, so don't worry about lost future ducklings. Her health is top priority.

When you introduce a new drake into your flock, there will be an adjustment period until he settles in. You may find many infertile eggs during this phase. Infertility can also be present with a young inexperienced drake, or with a drake that is too old. It is also dictated by the seasons.

Ducks normally commence laying at 21–26 weeks of age. Don't be concerned if you see a small amount of blood on the shell of the first egg. Though the first eggs are not as large as later ones, small blood vessels can rupture under the pressure of the first egg passing between the laying bones at the rear, or vent, of the duck.

You will notice that the duck's bill will become lighter in colour when she is in full lay, and her undercarriage will be heavier. This is quite normal.

Egg production is seasonal. Ducks require daily exposure to 14 hours of daylight for 3 weeks before they will produce eggs. Artificial light from a 60-watt incandescent globe hung in the duck house can be used to supplement daylight and bring the ducks into lay early.

To stimulate mating, drakes should be exposed to artificial light at least 2 weeks before the ducks, as the drakes take longer to come into season.

Remember, it takes time for sperm to swim into the oviduct, and it can be 7–10 days after mating before fertile eggs are produced. (Inversely, if you want to avoid fertilised eggs, you will have to wait 10 days after the drake is removed before the duck begins to lay infertile eggs.)

When breeding ducklings, always hold some as backup stock and new parent breeding stock. These young ducks will commence laying around 26 weeks of age. Often this is in the 'off season', especially if you have retained ducks from the first hatching. The eggs of young ducks will initially be small and infertile and of no value to the incubator.

Handling ducks

Ducks must be handled gently as their wings and feet are easily injured. Never grasp a duck by the wings or legs; grasp them by the neck or use a net, then place one hand over each wing to subdue them. Slide your hand beneath the breast as you pick them up and secure the legs in your hand so the bird's weight is resting on your forearm.

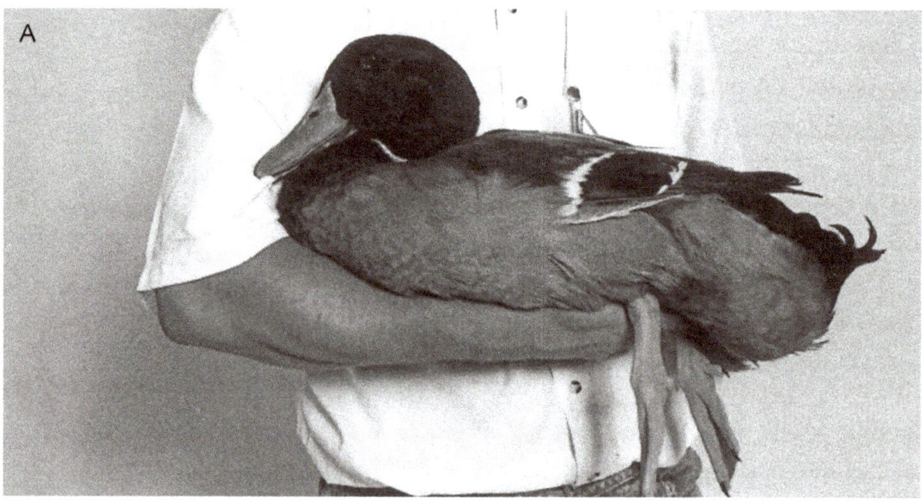

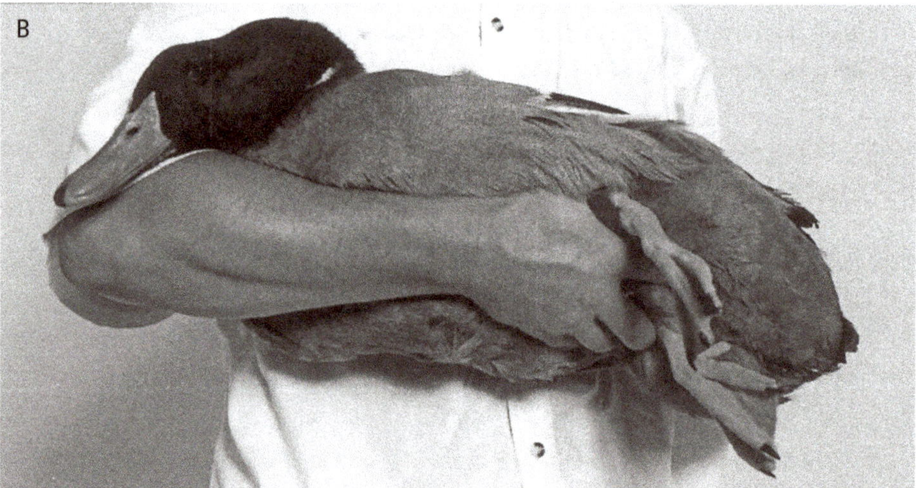

Figure 7.2: (A) The correct way to hold a duck. This Rouen drake is comfortable and secured. (B) Showing placement of the fingers. One leg is secured by the thumb, and the other by the middle finger. The palm of the hand supports the abdomen, while the forearm supports the keel and breast.

With the other hand, gently feel either side of the vent for two pointed bones. These are laying or pelvic bones. The distance between them will depend on whether a duck is laying. They will be approximately 2.5 cm apart on a duck that is not laying and as much as 7.5 cm apart on a laying bird. In some breeds you will notice that the duck's bill changes colour when she begins to lay. For example, in a Pekin the bright yellow bill will become very pale yellow.

Sexing ducks

Sexing ducklings

The only sure way to sex ducklings is vent sexing. In fact all waterbirds are alike and are very easy to sex. However, it takes gentle practice and it is best if someone with experience first demonstrates the technique for you. The procedure also requires understanding of duck physiology. Young ducklings are fragile and can be killed or permanently damaged by an inexperienced sexer. It is quite easy through rough handling to rupture the egg sac inside a young bird. It is best to practice on a duckling that is around 3 weeks of age.

The various steps are:

(1) Sex the duckling in sunlight if possible, as their organs are best identified in good light.

(2) Hold the bird with its back in the palm of your hand, with its head and neck resting on your wrist.

(3) Bend the tail of the bird backwards so that it is almost touching its back. Then secure the tail in that position with your index finger.

(4) Push the feathers away from the vent, and place your thumb in a direct line with the vent on the abdomen and gently press down. This will evacuate the bowel.

(5) Place your thumbs on either side of the vent and gently press down and slightly apart.

(6) Using your index finger, apply some pressure down and out on the back side of the vent. This exposes the genitals. A young drake will expose a short pencil (penis) that will be in the middle of the cloaca. Examine it carefully. They are often almost transparent but they are

there. In some coloured birds, such as a Black Indian Runner, the pencil will be black and quite easily seen. A young duck will expose no pencil, just lots of wrinkly pink folds of skin.

Your speed and accuracy is important and will improve with practice.

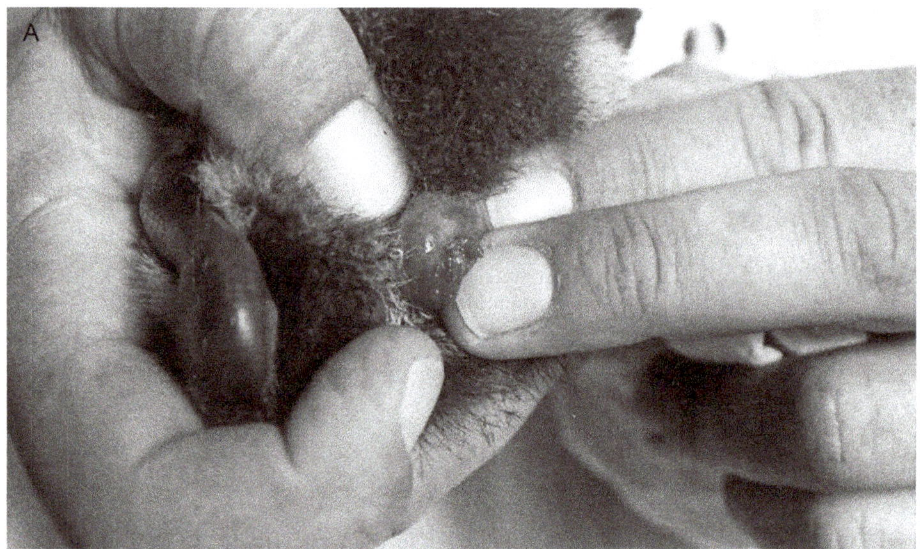

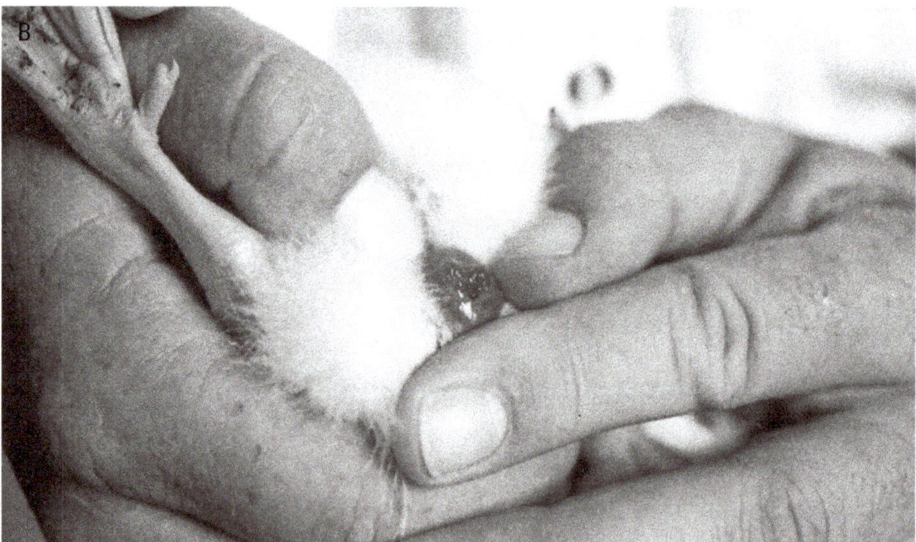

Figure 7.3: (A) Sexing a young drake. Notice the tiny pencil (penis) near the fingernail of the forefinger on the right. (B) Sexing a young duck (hen bird). Obviously there is no pencil.

Sexing older ducks

Sexing older ducks is quite simple because, as they grow older, the hens will quack, and the drakes will have a muffled and quieter quack and three curled feathers in their tails. When sexing birds, be as gentle and quick as you can to avoid injury to their genitalia.

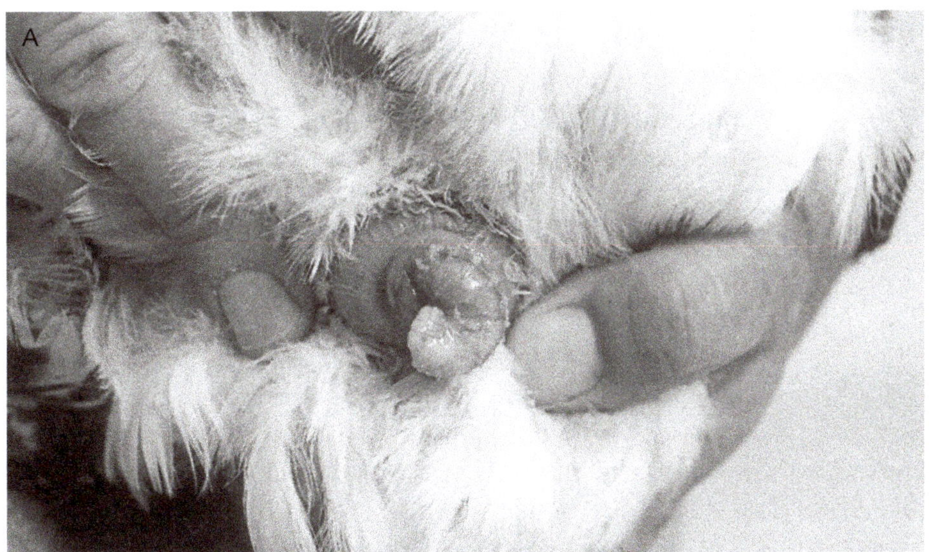

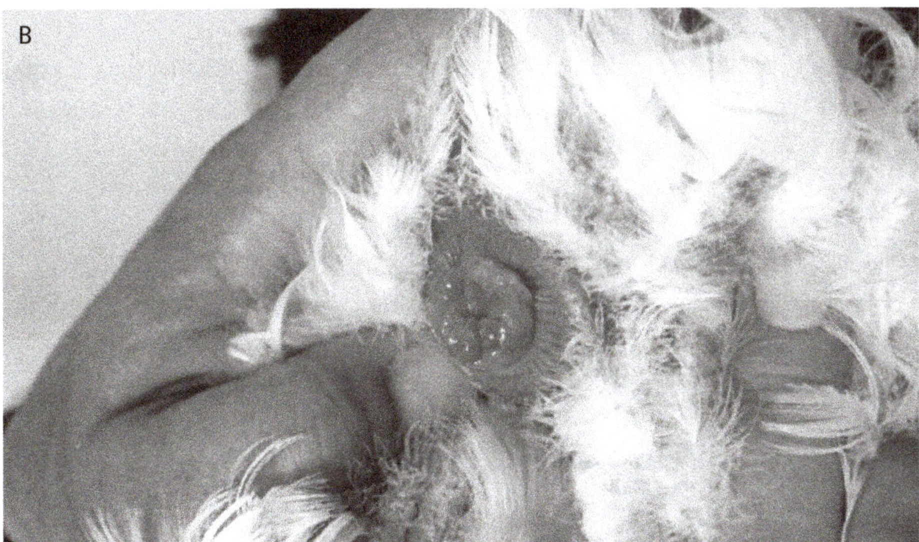

Figure 7.4: (A) The pencil of a drake. This matures in most breeds at 8–9 months of age. Until then, the pencil of a drake remains very small. (B) The genitalia of a duck.

Worming ducks

Worming waterfowl can be awkward for many reasons; the main one being they love to swim in the container and can overdose by drinking too much worming mixture. I have found that waterfowl ingest chemicals, good or bad, very easily. Not all wormers agree with waterfowl, so be sure that the one you use is suitable and effective. I have used a particular wormer on ducks and had some die.

Stagnant water increases the chances of your birds contracting worms, and worms are more prolific in the spring and at the beginning of autumn. This is when grass is warm and wet. An infestation of worms can retard growth and lower egg production. A serious infestation can kill your bird. Worms are also responsible for many diseases and illnesses.

Birds can be wormed once or twice a year. Piperazine solution is a good, safe wormer. Both it and Nilverm Pig and Poultry Wormer are available at stockfeed and rural outlets. I don't dilute the wormer with water. It has been my experience that worms are expelled alive when the wormer is diluted. I administer the wormer orally, using a syringe, at a dosage of 1 mL for every 2 kg live weight for any duck over 10 weeks.

I worm at night when the birds are relaxed, catching them individually and administering the wormer in the following manner:

(1) Hold the bird's bill open and pull the tongue forward to close the air passage. This prevents the wormer squirting down the windpipe.

(2) Squirt the wormer down the back of the duck's throat.

(3) Place the treated duck in a pen and then catch the next one until they have all been done.

I find this method very effective in that I know that every bird has been treated. They can't swim in the wormer and because it is not diluted, the worms that are passed are dead on the shed floor, not all over the yard.

I have tried garlic as a wormer and personally don't hold much faith in it. I found that when I placed garlic in the water dishes, some ducks ate it, so there was none left in the water. Placing the garlic in a stocking hanging in the water is not a solution, as it goes bad very quickly. I have concluded that using garlic is time-consuming and to be effective it must be in the water all year round. Now that's a lot of garlic!

As a rule, worms should not be a huge problem with your ducks. The digestive system of a duck is all downhill, and nothing stays in them long enough to enjoy the view or take hold.

Stopping ducks from flying

There are a couple of options for keeping the little webbed feet of smaller breeds on the ground:

(1) they can be pinioned

(2) the primaries can be plucked or cut.

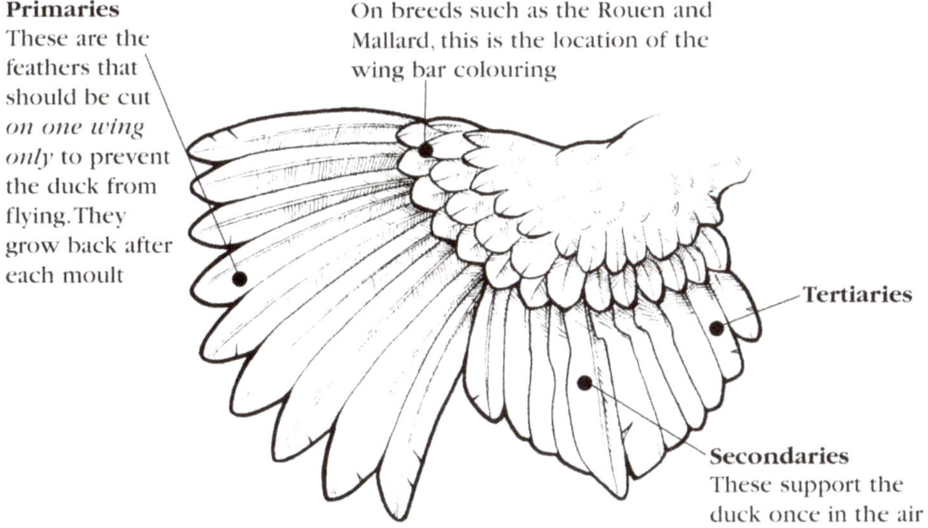

Figure 7.5: The wing of a duck.

The primaries are the wing feathers that are most important in pushing a bird forward in flight. They are attached to a single bone that corresponds to the first and second fingers of a hand and the fused hand bones.

The secondaries are the feathers of the inner wing. They are attached to what corresponds to the lower arm bone, and play an important part in supporting the bird in the air. The primaries and secondaries can be used separately. Tertiary feathers are attached to the upper arm bone. Each wing feather overlaps the one next to it, starting from the base of the wing outward. On the downstroke, the air pressure on the underside of the wing forces these feathers into an airtight fan. Speed and forward motion are gained. On the upstroke, the wrist joint is bent and all the primaries and secondaries turn on edge, like the slats of a Venetian blind, allowing the wing to lift with the least wind resistance. Slow-motion pictures show that as the wings move downward, forward, then quickly upward, their tips move through a figure-of-eight pattern.

Pinioning

Some people pinion bantam breeds, Mallards and some breeds of geese to prevent them flying away. This involves severing the tip of *one* wing of day-old ducklings. By removing the wing tip, the primaries do not grow and the bird will never fly.

In such young birds, the blood congeals easily. Any pinioning of adult birds should be done by a veterinary surgeon, as birds will go into shock and

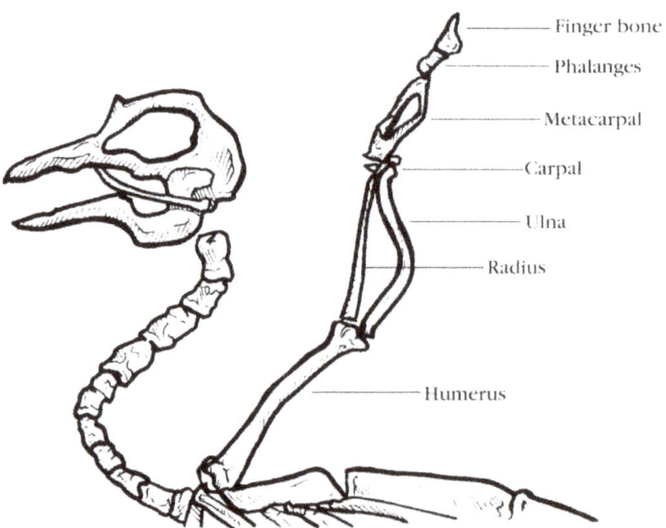

Figure 7.6: The wing bones of a duck. The last three bones house the flight feathers and are the bones that are amputated when pinioning.

quite possibly die from loss of blood. Pinioning is not a practice I support. It maims the bird for life and there are less destructive alternatives; however, having stated my personal view, I will explain how it is done.

A duckling's wing is made up of seven bones. The large bone that leaves the body is the humerus. It is joined to the ulna and the radius. Moving further out, these are joined to the metacarpal, from which the carpal protrudes. The metacarpal joins the phalanges bone and a small finger bone at the end. These last three bones underlie the flight feathers, and it is these that are amputated, using very sharp snips, at the day-old duckling stage (see Figure 7.6).

Dip the bleeding wing tip in caster sugar to promote clotting and separate pinioned ducklings from the others, who may peck at the blood and prevent the wing from healing.

Alternatives to pinioning

There are two alternatives to pinioning that you can use to stop your ducks from flying off:

(1) By cutting the flight feathers of one wing with a pair of scissors, the bird will be rendered flightless until the following moult. Obviously you will need to trim the flight feathers after each moult. Once the primaries have been cut, a bird will often choose to remain in the yard, provided it is content with its environment.

(2) The flight feathers can also be plucked, but this is not viable, as the missing feathers will immediately start to grow and in 6–8 weeks the bird will once again be able to fly.

I don't cut wings or pinion. My birds are mostly free to range at will. I would rather they be able to fly from a predator than be caught. Besides, flying away has never been a primary concern for me. I find it is far more common for birds to fly in.

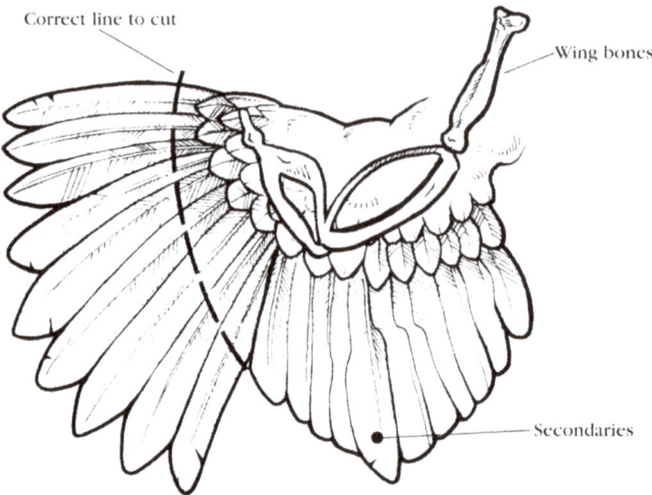

Figure 7.7: Cutting flight feathers. When clipping a wing it is not necessary to clip the secondaries as they are not responsible for lifting the duck into the air. Before you clip with the scissors, find the wing bones with your fingertips to make sure you don't cut them.

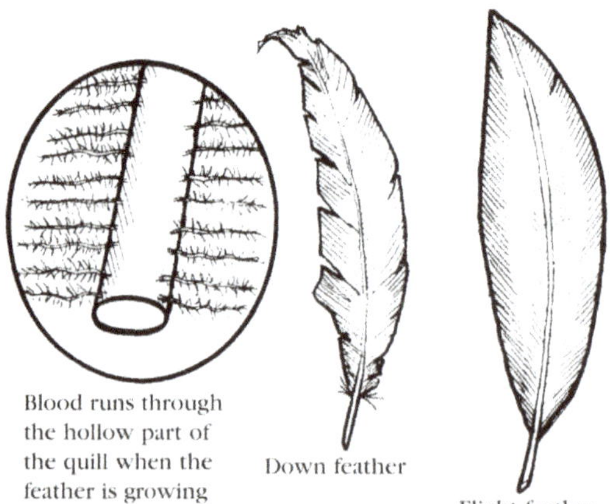

Figure 7.8: A flight feather, a down feather and an enlargement of a hollow quill. The flight, or primary feather, has a tough shaft and dense webbing. It is placed at an angle when the bird is flying and cuts through the air like a knife. For its lightness it has remarkable strength and flexibility. The down feather does not have a strong shaft. It is used in nesting and to keep the bird warm.

Some years ago I had a wild Mallard fly into my yard and lay an egg. We incubated this egg, not knowing it was a wild duck egg. When it hatched there was a household debate as to whether it was a Black Muscovy or a Pekin gone wrong. Of course when it grew up, there was no mistaking this duck for what she really was. She stayed in our yard running contently with the Pekins. She bred with them and the offspring were gloriously coloured in the wild Mallard colours, but had the body and wings of the Pekin, hence they could not fly. To this day we have this line of bird in our yard, and call them 'Garden' ducks. Most who see them call them Mallards, Rouen Claire and even Appleyards. I guess the moral to this little story is know and understand the breed you are buying.

Killing a duck

I have never killed a duck. I sell my culls at a dramatically reduced price. If you must kill a duck, make sure you do it humanely. The most humane way to kill a duck is by decapitation.

8
Duck housing

A fox-proof enclosure

Your duck enclosure should be fox-proof as ducks are at their most vulnerable at night. The surrounding wire fence should be high and buried at least 15 cm into the ground and then angled outwards. This prevents the fox digging under the wire to gain access to your birds. A 1.8 m fence is suitable. Don't be fooled. Foxes *can* climb fences. It is highly unlikely that they will climb your fence during daylight hours.

Figure 8.1: An enclosure with a 1.8 m fence. The yard is being rested to allow the grass to regrow.

Some people suggest that loose fencing is best, as foxes can't get a firm hold on a floppy fence. Our fences are loose, and South African ryegrass grows along the outside of the fence. This is a very thick and hardy weed that grows in a clump. It is almost impossible for anything to dig it out. On the inside of the fence line I have planted elephant grass, or banner grass, as it is also known. This is a broad-leaved grass with growth very similar to bamboo. It reaches a height of approximately 4.3 m and will become a living wall of green if left to its own devices. It also offers a little more privacy as a screen from the road. All of our poultry find this grass very tasty. Our geese keep the banner grass under control.

The night house

Ducks are said to be 'fully clothed', meaning they do not have a wattle and comb like a hen. They are more able to weather the elements; however, it is still necessary for them to have a house to go to at night that is dry and draught free, but well ventilated. This will also be the place that your ducks will lay their eggs. It is advisable for your duck house to have either a concrete floor or steel meshing on the floor to prevent any fox that successfully negotiates the perimeter fence from digging in during the night.

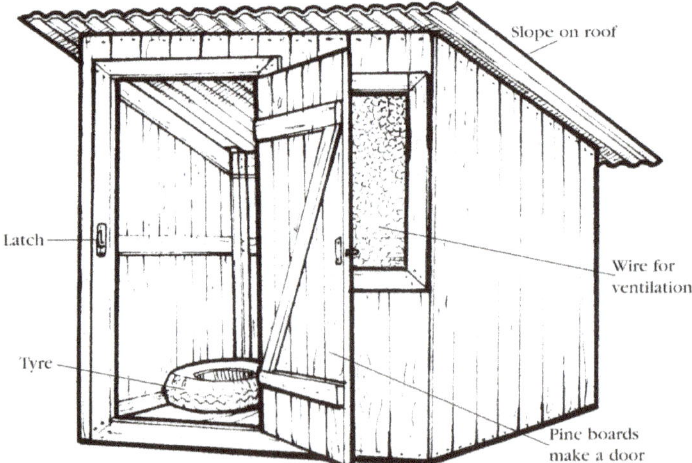

Figure 8.2: Duck housing. A night house like this can be made from recycled pine boards, corrugated iron (the roof), concrete slabs (the floor) and chicken wire (the window, for ventilation). Then all you need is sawdust shavings on the floor and tyres for the ducks to lay and nest in.

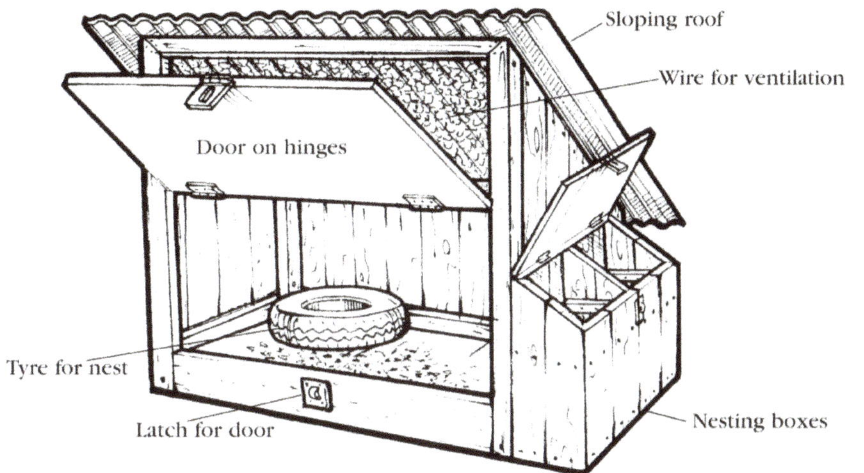

Figure 8.3: An alternative design for duck housing. The nesting boxes are optional.

A duck house doesn't have to be elaborate. Face the front of it away from the prevailing weather, using that side as a large window. Cover the window with steel mesh. You will need a roof that keeps the rain out and a door that can be closed and secured. The litter on the floor can be pine shavings or jarrah sawdust. I find that sawdust is economical. It is cleaner than straw, is absorbent, doesn't mat or smell, and reduces the infestation of flies.

Duck house equipment

The duck house needs to be furnished with feeders, waterers and nesting boxes.

I don't tip food onto the ground, but instead use large steel trays and pieces of guttering that I have sealed at either end to contain the food. This way the ducks shovel the pellets into their bills without the dirt.

Making a feeder from guttering is very simple and cost-effective. Cut the ends of the guttering with tin snips so you can fold them inwards. Pop rivet the overlapping pieces together so there are no sharp edges on your feeder.

A couple of tyres lying on their sides can be used as nesting boxes. They are ideal when your duck goes broody. Her eggs don't roll away, and when the ducklings hatch, there is little chance that they will wander out from under her at night and die in the cold. Even an empty 200-L drum cut in half or open-ended will suffice. These drums are great for geese to nest in as well.

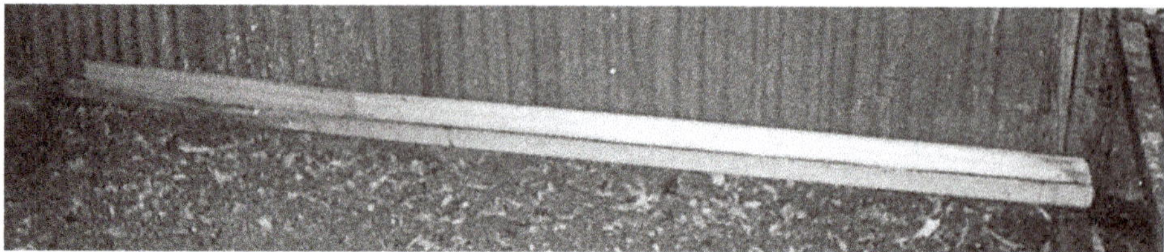

Figure 8.4 (top): Galvanised guttering used as a feeding tray. This prevents wastage and keeps the food clean.

Figure 8.5 (left): Half a 20-L plastic drum also makes a good feeder.

The duck yard

The duck yard should have some grass for the ducks to pick at. This is also conducive to insects and other bugs that ducks like to eat. A suitable and hardy grass is kikuyu; this stuff is just about indestructible. I have been running my waterfowl and poultry on kikuyu for 10 years and our yards still look like a bowling green. All the birds eat and pick at the new blades of grass, and each year I throw some fresh clean sand over the grass and fill in any areas that have been eroded by the ducks billing the ground.

Every week it is a good idea to hose down the grass, watering in any manure that is on the surface. This will inhibit flies.

I have two main yards that I run the birds in. While one yard is in use, the other is having its yearly rest. I use this resting period to plant new shrubs or grasses, as it is a waste of time planting anything while the birds are watching. The minute you turn your back, they pull it out.

Bamboo is a good grass to plant. It grows exceptionally well, provides shade and cover for your birds, requires little maintenance, and during summer you can place a sprinkler through the tall leaves and it acts as an air-conditioner with the breeze blowing through.

I find bamboo very appealing and use lots of it for screening yards from one another. Cocos palms add a very tropical flavour to our yard. The beauty

Figure 8.6: Ducks foraging in their enclosure. Grass is ideal as it attracts insects and frogs.

of the Cocos palms is that they allow light through, encouraging the grass to grow right up to their bases, so there are no balding patches beneath them.

Depending on the size of your enclosure, you may be able to plant eucalypts, such as blue gum, that will grow rapidly and provide a lot of shade. In winter it is best to cut the top out of these trees to encourage them to become more bushy than tall and uncontrollable.

The duck pond

It is important for your ducks to have a pond or bathtub in which to swim and preen. Remember, if they can't get out easily, you run the risk that they will become exhausted and drown. Baths recycled as ponds are one of the biggest killers of waterfowl. When the birds get into the tub and splash about, the water level drops, and then they can't get out. If you do have a bathtub as a pond, place some rocks in it to form steps suited to short duck legs.

Ducks filter water and sand through their bills when sifting through the mud for food. This prevents a build-up in their salt ducts and washes out their eyes.

Figure 8.7: The perfect duck pond. Notice that the pond has graded sides and that the shallow end drops away slowly – this makes it ideal for ducklings.

The heavier breeds of ducks, like the Aylesbury and Rouen, often mate in water, as their size makes mating an awkward exercise on dry land. This also achieves higher fertility, as the water lubricates the male genitalia and prevents male prolapse.

Remember that ponds are great for ducks, but may not be so good for young children. Be sure to take the appropriate safety precautions.

9
Duck diet

Ducks that range free in grass and have access to a pond will find some of their own food in the form of grass, insects, slugs and water fleas. They might survive on such a diet, as wild ducks do, but they will not lay well or fatten effectively for the table.

Nutritional requirements

The functions of a correct diet are to:

(1) rebuild tissue as it constantly wears out
(2) form new tissue, particularly in the case of growing stock
(3) produce new substances, such as eggs
(4) generate heat and energy.

Heat is generated by the slow oxidation of starch, sugar and fats; that is, the union of the oxygen carried by the red corpuscles of the blood with the carbon and hydrogen in these compounds. Since oils and fats contain carbon and hydrogen in greater proportion than do sugars and starch, the former generate more heat, weight for weight.

The special functions of nutrients such as carbohydrates, which are heat producers, include the fact that they can also serve as fat formers. Oils and fats produce heat and energy, the oily secretions for the joints and under the skin, and are laid up in the body as fat.

Nitrogen substances form the muscles, connective tissue, skin, feathers, the greater part of the secretive and excretive organs, and eggs! Ash is essential for bone formation and for some digestive juices. Fibre is usually considered of small value, but in certain foods a considerable portion of this is digested.

Vitamins

Vitamins play a very important part in your ducks' diet.

Vitamin A is fat-soluble and readily destroyed upon exposure to heat, light or air. It plays a direct role in vision and is a component of a pigment present in the retina of the eye. It is essential for the proper functioning of most body organs and also affects the functioning of the immune system. Growing ducklings, especially, need vitamin A.

The vitamin B complex consists of several vitamins that are grouped together because of the loose similarities in their properties, distribution in natural sources and physiological functions. All the B-group vitamins are soluble in water. Most of them have been recognised as coenzymes, and they all appear to be essential in facilitating the metabolic processes of all forms of animal life. The B complex vitamins include B1 (thiamine), B2 (riboflavin), B3 (niacin or nicotinic acid), B6 (a group of related pyridines), B12 (cyanocobalamin), folic acid, pantothenic acid and biotin.

Vitamin B1 helps ducks convert carbohydrates into energy and helps in the metabolism of proteins and fats. Vitamin B2 is required to complete several reactions in the energy cycle. Vitamin B12 is a complex crystalline compound that functions in all cells, but especially in those of the gastrointestinal tract, the nervous system and bone marrow. It is known to aid in the development of red blood cells in higher animals.

Other important vitamins include D, E and K. Vitamin D is a fat-soluble compound essential for calcium metabolism in animals and, therefore, is important for normal mineralisation of bone and cartilage. Vitamin E is a fat-soluble compound. The metabolic roles of this vitamin are poorly understood. Its primary role appears to be as an inhibitor of oxidation processes in body tissues. Vitamin K is fat-soluble and essential for the synthesis of certain proteins necessary for the clotting of blood.

Minerals

Minerals are non-organic. Most foods contribute to mineral intake. Spinach, for example, is high in iron, which builds up the red blood cells.

Calcium is essential in all birds' diets. It contributes to their early development and to the shell of the eggs they lay. Calcium can be provided as limestone, coral lime, or as a powder added to feed. Phosphorus goes hand in hand with calcium for building strong bones.

Minerals are not something we spend a great deal of time on when dealing with waterbirds, because ducks are continuously billing the ground, eating grass, bugs and sand. In doing so they find all the minerals they require, providing they are not being run on concrete.

River sand and limestone make good digestive materials for them, and are suited to both ducks and geese. These materials assist the birds' digestive organs and the consumption of them is quite normal. Limestone is obtained from shells and corals and is high in calcium carbonate. A goose or duck that has been sitting for some time will often look for some limestone to eat when she comes out for a drink and a swim. It also neutralises soil acids and prevents ducks from laying soft-shelled eggs.

Ducks can be ferocious eaters at the best of times, but especially if their diet is lacking. This is why the spot you choose to run your ducks must be conducive to insects, bugs, frogs and other things that ducks like to eat.

Grass plays a very important part in providing vitamin A in a duck's diet. Lack of vitamin A will cause ducks' eyes to become weepy and sticky, which can predispose them to eye infections.

Feeding your ducks wheat and vegetable scraps does not constitute an adequate diet. I have found that a good diet consists of barley, wheat, maize/corn (one of the few grains that makes a duck's digestive system really work), small shell grit and pig pellets. Yes, pig pellets!

I feed pig finisher to my ducks. I have tried everything from making up different recipes from wheat, lucerne pellets, laying pellets, crushed maize, budgie shell grit and soya bean meal to buying premixed bags of grains. Unfortunately, I have found the quality of the bagged foods to be too inconsistent. One day it's great, the next the ducks won't touch it because the producer has changed the recipe or the bag was the last one out of the hopper and is mostly dust or just wheat. Wheat alone is not a substantial food.

Pig pellets are one of the few feeds that contain fishmeal and bloodmeal. I guess the proof is in the pudding. A duck I had for some time that had never laid an egg is now up there among the egg producers.

I am not one to cut corners on the feed I give my ducks, and am always reading labels to see what food preparations contain. Mostly I look for the ingredients, then the protein level. Ducks really don't need anything much over 16% protein.

How much do you feed a duck?

One of the most frequently asked questions by those new to keeping ducks is: how much food do ducks need? Most books recommend tipping sufficient food into a container to keep the birds eating for 5 minutes. Any more is judged to be excess that will deteriorate during the course of the day, as well as attracting unwanted flies. However, I have found that if I place plenty of food in the ducks' trays of a morning, they don't eat as much as when I limit their feed.

I feed twice a day: in the morning and just before I put the birds to bed in the evening. There is always food in the trays for them, and my food costs are not excessive. Whereas if I follow the recommended 5 minutes a day of feeding, they eat a great deal more in the course of a week, and are very noisy. A hungry duck is often a noisy duck.

Depending on its size, an adult duck will eat 170–200 g of food a day. When I place a new type of feed out for the ducks' breakfast, I watch to

Figure 9.1: Ducks, fowl and geese feeding together from a feeding tray.

make sure they are eating it. I don't believe in waiting for a bird to get so hungry that it will eat something that it doesn't like. If your birds are not eating, they are no longer egg producers or meat producers and you are the one who suffers in the long run.

Some breeders use stale bread to fatten their birds in preparation for showing. This ploy is used mainly with the heavier breeds, such as Rouen, Aylesbury and Muscovy. While it gives them a more desirable and impressive look at the show, an overweight duck is not a good breeder, and excessive bread can create digestive problems.

Don't place the food directly onto the ground. Not only is this practice wasteful, it can also pose a health risk to your birds. If it becomes moist and unpalatable, spoiled food opens the door to botulism. Rotting organic matter makes for ideal botulism conditions, and is always a concern with waterfowl.

Vegetable source feeds

Various cereal grains, consisting of carbohydrates and proteins, are the most important duck food.

- **Maize/corn:** is especially high in carbohydrates and fats. It should never make up more than 15% of the diet. In winter, it is also good for ornamental ducks and geese that spend a lot of time on the water. It is a warming food due to the high level of carbohydrates, and can colour feathers a lemon shade if birds are fed extra corn during their moult.
- **Wheat:** is supposed to be the most enriched protein grain, but I have never found any bird that will lay on wheat alone. If ducks are given too much wheat, the protein/carbohydrate balance in their diet may become upset and the food will not be used optimally. In other words, you will find whole grains of wheat in their manure.
- **Barley:** is not very popular with ducks, because it is a hard, pointed grain, but it has a positive effect on muscle production and should always form part of their diet. Some pelletised food consists of barley.
- **Flaked oats:** are high in fibre (10%) and less high in protein than wheat. They serve as roughage and have a beneficial effect on various bodily processes. It has been asserted that flaked oats will discourage feather

eating. Flaked oats also are accredited with a better breeding result. They are probably best prepared as a mash for ducks to eat.

- **Mill mix:** is a by-product of flour milling, which removes the outer cellulose walls of the kernels, together with bits of the sticky layers underneath, which are high in protein. Hence it is high in protein and a valuable duck food.
- **Stale bread:** is often given to ducks, is high in carbohydrates and is easily digested by the birds. Sometimes it is mixed with commercial feeds to complement the protein. Stale bread is often softened in liquid before it is fed to ducks.
- **Spinach:** can be chopped and steamed for ducks. It is very high in vitamin K.
- **Greens:** are loved by ducks. Grass doesn't have much food value, but it does contain vitamins and trace elements, and provides the roughage essential to the digestive process.
- **Soya beans:** are nearly one-third protein – about twice the level of protein in meat.

Animal source feeds

- **Meatmeal:** is protein rich (as high as 86–89%); however, it often has acids added to it, in which case it is not well suited to ducks. Too much meatmeal can cause slipped wing and health problems. If left in the sun, meatmeal can become a source of botulism.
- **Fishmeal:** is made from whole fish and fish by-products. It contains proteins that are essential in the diet of ducks, but is almost impossible to buy in Australia. Instead I buy pig pellets, which contain fishmeal. Cod-liver oil as part of the diet can be equally effective.
- **Milk products:** have easily digestible protein and a good source of vitamins. They can be fed in the form of cottage cheese, skim milk or yoghurt. Denkavit, a powdered milk, can be mixed in the feed in the morning if consumed at one feeding so there is no time for it to turn sour and rot in the sun. Skim milk can also be fed. Milk products are not something you would continually feed your birds.

Lime: should also be readily available to your birds as part of their feed or left in a container for them to consume at will. Coral lime is especially good if you have a duck laying soft-shelled eggs, or eggs that have a calcium-like tail on one pole. During the laying season ducks have an increased need for calcium.

Soft food: can be created by mixing food items, such as boiled potatoes with cereal meals and chopped greens, into a moist but not wet form.

Dietary deficiencies

Malnutrition

Malnutrition is indicated by stunted growth, retarded feather development, leg weakness, emaciation, and low resistance to diseases and parasites. Malnourished ducklings may be unbalanced and continually flip over backwards. A staggering walk can be followed by paralysis in severe cases. Small white pustules may be present on the lining of the throat and gullet, together with cheesy material in the bursa or fabricus. (The bursa is a round gland the size of a pea found dorsal to the rectum. Its function, like that of a tonsil, is antibody production.)

Cause: an incorrect diet, including a diet solely of grains, or insufficient feed and grassless yards.

Treatment: provide a balanced diet. Do not try to save money by starving ducklings or using cheap feed that doesn't supply a complete diet. If you are feeding grains, ensure that the birds have some access to insects and grass. If you are not prepared to do this, don't keep ducks!

Rickets

Ducks suffering from rickets fail to develop normally because of failure of the cartilaginous skeleton to become ossified (bone).

Cause: a number of factors contribute to rickets; the most common being a deficiency in vitamin D3. Other causes are calcium and phosphorus deficiencies and an imbalance between these two minerals and vitamin D3. Vitamin D can be synthesised by the body from ultraviolet light (the sun's rays); however, for this to occur, ducks need to spend time outside in the sun. Zinc in excessive levels is known to interfere with calcium absorption, and high levels of fluorine in rock phosphate can also predispose a bird to a soft bone condition.

Vitamin A deficiency

Vitamin A is essential for healthy growth. Deficiency results in various disorders that most commonly involve the eye, the skin and the mucous membranes lining the internal body surfaces, but may also result in defective bone formation and reduce the hatchability of eggs. Ducklings that do hatch are usually short-lived. An early symptom of vitamin A deficiency is the development of weeping eyes with a 'wet' sticky discharge. The cornea of the eye can become opaque, giving a whitish appearance.

Treatment: vitamin A supplements can be administered as a remedy together with vitamin D3, cod-liver oil, maize and fresh greens. In older ducks, the characteristic eye disease may be followed by severe lethargy and difficulty in balancing to the point where they cannot walk. In cases where paralysis has been evident for some time, treatment with vitamin A does not help, as the nerves supplying the affected organs are damaged beyond repair. These birds should be disposed of humanely.

Treatment of eye lesions in individual birds can be carried out by spraying an antibiotic, such as Terramycin, in each eye.

As incredible as it sounds, I have seen ducks at shows with these symptoms. It makes me wonder how much those showing them care about the bird.

Note that excessive intake of vitamin A causes a toxic condition. The symptoms may include bone pain, fatigue and drowsiness. There may also be blurred vision, growth failure, enlargement of the liver and nervousness.

Vitamin B complex deficiencies

Vitamin B1 deficiency affects the functioning of the gastrointestinal, cardiovascular and peripheral nervous systems, and results in weakness of the limbs. A typical sign of this deficiency is a duck that won't walk. Inflammation of the tongue, and a greasy, scaly inflammation of the legs are common symptoms of deficiency of vitamin B2 (riboflavin). Deficiency of vitamin B12 most commonly results in pernicious anaemia (a blood disorder).

Niacin, or nicotinic acid, helps the metabolism of carbohydrates. Prolonged deprivation leads to pellagra; a disease characterised by skin lesions, gastrointestinal disturbance and nervous symptoms such as twitching or head swaying. Ducks displaying signs of niacin deficiency can usually be cured by the immediate addition of a niacin supplement to their feed or drinking water.

If you suspect a duck is lacking B-group vitamins, add cod-liver oil to a mash. I have also given a macro B capsule dissolved in water to a duck for a period of 3 days with excellent results. I view this vitamin as a tonic for ducks that seem a little 'off colour'.

Vitamin D deficiency

Lack of vitamin D may result in rickets, bowlegs and turned-in feet. Thin-shelled eggs and low hatchability are other signs. Once again, cod-liver oil, which contains vitamins A and D, can be frequently added to mash given to ducklings. A large quantity is not required, the usual addition is 1–2% of the total feed. Vitamins A and D are not stored in the body, so cater for a daily allowance.

Vitamin E deficiency

Deficiency of vitamin E is rare, but may impair neuromuscular function. Should you purchase or be in possession of an overweight bird that you wish to breed from, it is advisable to place the bird on a strict diet to shed the excess fat, and slip a vitamin E capsule down its throat. Your bird will grow extremely amorous. This advice comes hot off the press from David Green in Queensland, a breeder of some of the best Toulouse geese I have had the pleasure to admire and purchase.

Vitamin K deficiency

Vitamin K deficiency, though relatively uncommon, results in impaired clotting of the blood and internal bleeding. Once again, if your duck is listless, it can be due to a lack of leafy green vegetables, in which case, introduce your duck to the garden. It is also worth knowing that if your bird accidentally eats snail pellets or rat poison, vitamin K is the antidote.

An overall tonic for ducks

Stockgain is a commercial product that resembles molasses in texture and smell, and contains far too many minerals and vitamins to list here. It is added to water to supplement the diet of horses, cattle and other stock. It can also be used to supplement your ducks' diet.

Stockgain can be administered daily. Mix 10 mL in 2 L of drinking water and allow your ducks to access the mixture on an ad lib basis.

10
Duck health

In the backyard situation, ducks rarely become ill. Most of their problems are diet-related or caused by stagnant water, or wormers or medications that are unsuitable for waterfowl. Sometimes a duck will have been bitten by a snake or stung by a bee, spider or ant. When a problem does appear, it helps if you are able to diagnose the problem and remedy it at the source.

You don't really have a lot of choices when it comes to advice about ailing ducks: you can consult a duck breeder or a veterinary surgeon.

I am not being cynical or sarcastic when I say I have felt myself surrounded by a huge brick wall when I have sought treatment for a sick bird. So the advice I offer here is what *I* would do. I am *not* a vet – just someone who loves his ducks.

Prevention is, of course, better than cure. Sunlight is the best purifier and all poultry sheds should be well lit and well ventilated, with large windows. Sanitation is of utmost importance, as some diseases are contagious.

Sick or injured ducks should be isolated in a warm, dry, dark, draught-free place. Ducks frighten easily and go into shock when they are unwell. You will know a duck is unwell if it lies on the ground, is reluctant to go into water, the breastbone protrudes, or the feathers around the abdominal area are dirty and look stringy. In sick birds the natural reaction to move away from you is not present, and they show no interest in food.

If you find a bird you believe is unwell, begin by assessing the symptoms. You may find the following summaries useful in diagnosing and treating a number of duck maladies.

Duck diseases

The following diseases are contracted as a result of poor sanitation and gross mismanagement of ducks. I would be astonished if you needed to refer to this short list. In writing a detailed book on ducks it is necessary to list all information; however, in Australia ducks are relatively disease free. This is something for which we can be grateful to the Australian Quarantine and Inspection Service (AQIS).

The diseases mentioned here are all 'quarantinable' and the department of primary industry or agriculture in your state must be notified of any outbreak.

Aspergillosis

Aspergillosis is a disease that affects many species of wild and domestic birds and is caused by the fungus *Aspergillus fumigatus.* It usually affects ducklings in the first 2 weeks of life. Small ducklings may die suddenly without symptoms having been evident. Laboured breathing and other respiratory symptoms are usually observed in older ducklings. There is no treatment of value.

Cause: mouldy straw appears to be the main contributing factor, so special care should be given to the selection of bedding for ducklings.

The spores of the aspergillosis fungus become airborne at the slightest movement and are inhaled by the ducklings. In healthy birds the spores remain inactive, but in ducklings with lowered resistance due to a cold and draughty environment, a lining of fungi forms on the moist inside walls of the respiratory tract and organs, clogging the narrow passages. Aspergillosis is not transmitted from sick ducks to healthy ones.

Prevention: ensure that the housing is well disinfected and that incubators are regularly fumigated.

Duck cholera

Duck cholera is a contagious disease to which domestic poultry, chickens, ducks, geese, guinea fowl, turkeys, pigeons, pheasants and fancy birds are susceptible. It occurs all over Europe, in North and South America, and most parts of Africa and Asia. It is characterised by the onset of sudden fever, diarrhoea and extensive diffusion of blood into various organs. It is usually fatal, with birds collapsing and dying 2–4 days after onset of the disease. Affected birds droop their wings, refuse food and water, and there is great nervous prostration, rapid respiration, a discharge from the eyes and nose, frothy saliva from the bill, and severe diarrhoea that soils the hind parts of the body.

Duck virus hepatitis

Virus hepatitis is a highly infectious disease of ducklings. Goslings are resistant. It is characterised by its sudden onset and rapid, violent course, and in ducklings less than 3 weeks old, up to 90% may be dead within 2 days of the first losses.

Resistant ducks can be bred and a vaccine has proven successful. This is a notifiable disease, so you must notify the department of primary industry or agriculture in your state of any outbreak. This disease should be suspect if a duck is found with its head stretched upwards and backwards. Only birds that are dead or dying show the accumulation of lipid in the convoluted tubules of the kidneys.

Salmonellosis

Salmonellosis can spread through your birds like a bushfire, generating huge losses, especially if ducks are bred on a large scale. The initial cause may be dirty eggshells, insufficiently fumigated incubators or brooders, or unsanitary conditions in general. Infected ducks will eliminate the bacteria in their droppings. Salmonellosis is transmittable to humans. It is also a notifiable disease, which means that you must inform the primary industry or agriculture department in your state of any outbreak as immediate quarantine is required.

Symptoms: the most common signs are diarrhoea, swollen joints and paralysis. Ducklings will look for warmth, with raised feathers and closed eyes. Death results. Ducks that survive this disease will retain the bacteria and pass it on through their droppings and eggs for the rest of their lives, which means that all must be destroyed humanely.

Septicaemia

Septicaemia is the likely culprit when a duck is found dead, having previously shown no sign of ill health. This is a condition in which bacteria are circulating in the bloodstream. It is very serious because the organisms and the products of their activity (toxins) become widely distributed throughout the body tissues, affecting practically every organ.

In most cases, septicaemia results in death from within a few hours of contracting the disease to 6 or 7 days later. Frequently, especially when the duck is in a weakened state, sudden death is preceded by a very high temperature. (Ducks' normal temperature is 42.8°C.) The high temperature may be the only sign of the presence of septicaemia, but other symptoms include loss of appetite, green diarrhoea, in some cases staggering, swollen hocks, and discharge from the eyes and nostrils. If weakened ducklings are disturbed while sitting, they will attempt to walk but fall over, sometimes showing uncontrollable nervous movement of the head and neck.

Treatment: antibiotics and/or sulphaquin. Read the directions on the medications and adjust to the body weight of the bird. Treating the rest of the flock with sulphadimidine in the water, or sulphaquinoxaline in the feed, can be effective. Erythromycin and chlortetracycline have been used with success against septicaemia.

Duck parasites

Lice and red mites

Lice and red mites can attack ducks. The birds should always be regularly inspected for signs of lice and/or mites. If infested, a poultry dusting powder may be applied, although I find having ample water for them to swim in soon eliminates this problem.

Gizzard worms

Echinuria worm (tapeworm) eggs enter the duck's food chain via the water flea. The eggs develop into larvae within a few days and find their way into the duck's stomach, burrowing deep into the lining, where they form cysts. These cysts block the true stomach and halt the progress of food, and at this stage the brain may also be affected.

Symptoms: ducklings to 8 weeks of age are most susceptible and will be disinterested in food, retching and shaking their heads vehemently. After a few days they become weak.

Treatment: Piperazine or Levamisole wormer. Ducklings that survive will continue to pass worm eggs as segments of the worm are expelled. These infestations can be prevented by restricting ducks to artificial ponds where there are no water fleas, rather than allowing them to swim in natural waterholes and streams. Once the ducks reach maturity, the worms no longer present a danger.

Stickfast fleas

The stickfast flea was first recorded in Western Australia at Geraldton in 1913. It became a serious problem in Perth in 1920; however, its incidence dropped when concrete floors were adopted in commercial poultry raising. Unfortunately, many backyard poultry breeders are still at the mercy of stickfast fleas.

The complete life cycle of the stickfast flea is about 4 weeks. The female lays her eggs at night while attached to the bird, and the eggs fall to the ground. In about 4 days the worm-like larva emerges. This feeds on organic material, mainly dried blood excreted by the adult fleas, and shelters on the surface dust and litter on the ground. It grows to about 3 mm, burrows down, spins a cocoon, develops into an adult flea and emerges to search for a host.

Young ducklings can suffer serious injury from stickfast fleas, with some dying from an infestation. The fleas bury their mouthparts into the flesh, breathe through the rear portion and remain fixed in position for the remainder of their life.

Treatment: smearing the infected parts of the duck with a greasy non-burning substance, such as Vaseline, lard, neat's-foot oil or vegetable oil, will control the adult fleas by suffocating them.

'Flyaway' is a totally natural product used to keep flies away from horses. It kills stickfast fleas instantly. The following day, they are no more than crusted shells that fall off the duck's face.

To prevent stickfast fleas, never bring an affected bird home. If it is already on your property, clean out the ducks' house, removing any loose equipment.

Clear away any debris in the yard, and burn the rubbish along with the litter from the duck house. Maldison (malathion) is a suitable poison with which to spray your shed. Read the directions carefully: 1 part Maldison with 17 parts water should be applied to any crevices or cracks. Make sure the ducks are outside and away from the poison while you spray.

Kerosene is a very cost-effective alternative way to eradicate stickfast fleas in the shed, but once again, don't spray kerosene on your ducks. If using this method, you will need to spray again 12 days later.

Ticks

Ticks are a common problem in many parts of Australia. Even though the symptoms are very straightforward, quite a few people overlook the obvious. When you have a lame duck or drake, ticks should be one of the first things you consider. Ticks will kill your birds, so you must take positive action immediately.

Ticks usually stay in the small crevices and cracks of the wood of duck sheds during daylight hours and come out at night to feed on your ducks. They suck the duck's blood and inject a toxin that eventually causes paralysis. Sometimes a tick will leave smaller ticks behind. These can be easily seen on the thighs and beneath the wings of affected birds as they look like little black sultanas.

Treatment: the best means of removal is to spray methylated spirits on them. By the next day they will have dried up and will brush off easily. Never pull ticks off your bird without using methylated spirits first; they will break in half leaving their head in the skin of your bird, which may result in blood poisoning and loss of the bird. If penicillin is injected into the leg muscle of a duck paralysed by the tick toxins, the bird will recover within 48 hours.

To effectively rid your shed or duck house of ticks, saturate all wooden surfaces with kerosene and lightly spray the bedding, making sure there are no flames or sparks nearby. This will halt the entire cycle of the tick. A point of interest: a tick can survive for 7 years in a glass jar with no food.

Worms

Worm infestations can occur in ducks. Administering piperazine citrate through drinking water may treat bad cases. Adequate vitamin A levels assist in maintaining resistance to worm attack. A reputable duck breeder has suggested placing a small amount of Condy's crystals in the ducks' drinking water once a week. This places a lining on the gut wall and prevents any parasites sticking to it.

Other common health problems

With correct care, diet and management, you will very rarely have a sick duck. And I hope you need never refer to these few pages.

Botulism

Botulism is food poisoning and can be deadly. It is a toxin found mostly in decaying animal and plant matter. Even rotted food from household scraps can cause botulism. It flourishes when the weather has been dry and water levels in ponds or lakes drop, leaving decaying plant and animal matter exposed where ducks can eat them. It cannot survive in aerated water.

Ducks in suburban parks and public gardens frequently suffer and die from botulism. The most important first step is to remove any dead carcasses and burn them. If you bury them, botulism can be transferred to any creature that digs them up.

A few hours after eating poisoned food, ducks will lose control of their leg, wing and neck muscles. Ducks that are in the water will drown when their heads flop into the water. Birds on land will slip into a coma before dying. Botulism can kill within 3–24 hours.

Treatment: take one-quarter teaspoon of sulphaquin and tube this raw down the throat of the duck. Place the same amount of sulphaquin in the duck's drinking water for 4 days. This treatment was shared with me by Dave Green of Queensland on an occasion when I assisted in treating birds suffering from botulism. At the time we were tubing the birds with castor oil, Epsom salts, Glucodin, Pentavite and Terramycin powder. This treatment involves the following steps:

(1) Mix half a teaspoon of Epsom salts into half a cup of water for the first treatment.

(2) Check in the bird's bill. If there is mucous on the tongue or in the throat, remove it with a cotton bud.

(3) Take a piece of small plastic tubing and apply cooking oil or molasses to the outside. Open the duck's bill and pull the tongue forward, so you don't push the tubing down the windpipe.
(4) Gently slide the tubing down the back of the duck's throat and into the crop, while holding the bird's head upright. Using a syringe, inject 10–25 mL of the mixture into the duck via the tubing.
(5) Remove the tubing gently, and hold the duck's head upright to prevent welling.
(6) Repeat this treatment an hour later, and every following hour until you see a bowel movement.
(7) The final dose of Epsom salts mixture is followed by a maintenance mixture of half a teaspoon of Terramycin powder dissolved in half a cup of water, together with 1 teaspoon of glucose or molasses and a few drops of Pentavite or Stockgain. Administer this using the tube and syringe in the same way as the Epsom salts doses were given.

Make up fresh maintenance mixture each day, in doses of 10–20 mL, depending on the size of the bird. When the bird improves, add 1 teaspoon of Farex to the mix. Cod-liver oil is also beneficial.

Colds and chills

Ducks and ducklings can contract a cold. When this happens, a watery discharge will run from the nostrils and eyes.

Prevention: always ensure that the ducks' bedding is warm and their housing free of draughts. It is particularly important that ducklings are warm at night. Ensure they don't become excessively waterlogged during daylight hours. If the weather outside is lousy, wet and windy, leave the ducklings under a brooder light.

Treatment: wash the eyes and nostrils of affected birds twice a day with Condy's crystals and place a small amount of Condy's crystals in their drinking water – just enough to make the water go pink. Small amounts of Terramycin can also be used in the drinking water.

Eversion of the oviduct

Eversion of the oviduct may occur when a duck has been straining to lay an unusually large egg, or when she has been overworked by the drakes. The oviduct muscles are weakened and the oviduct is expelled. The prolapsed oviduct appears as a dark red swelling protruding from the vent. Other birds can be attracted by the swelling and peck at it, frequently leading to death of the duck. If these ailing ducks are left unattended, they will die a slow and painful death.

Treatment: a duck can be saved if the eversion is noticed within a few hours of it happening, although chances of healing vary.

Begin by carefully checking that there isn't an egg in the duck's abdominal cavity that might have caused the problem. If there is, inject 10 mL of warm vegetable oil or paraffin oil into the vent, and gently massage the egg along the oviduct until one of the poles (ends) appears. Then slide the egg out. If the shell breaks, you must flush out the fragments so they do not rupture the oviduct wall.

Wash the oviduct thoroughly and gently with warm water and disinfectant (a mild solution of Dettol or something soothing). Applying heaps of antiseptic ointment so the oviduct is slippery, gently try to reinsert the prolapsed organs into the abdominal cavity. Do not use force. Speak softly to your duck to calm her so she will relax.

When successful, isolate her in a dry, warm place for a week or so to allow the muscles time to heal. Feed her light grain-free meals of leafy green vegetables with small amounts of bread, mill mix and a teaspoon of cod-liver oil. Molasses, Denkavit and ground limestone are other suitable additives to her diet.

Dislocated neck

When transporting mature birds, such as Indian Runners, make sure the boxes they are transported in are not so low as to force the duck to hunch up, as this can throw the neck joints out. When this occurs, the duck won't be able to stand upright on arrival and will resemble a hunchback. (It is similar to us bending down for an extended period of time and having our back lock. It is an extremely painful and uncomfortable experience.) This problem is easily avoided by having a suitable crate in which to transport ducks.

Treatment: involves gently and carefully realigning the vertebrae. Hold the duck with both hands in front of you, pointing its head and neck away from you. With your thumbs, start at the base of the duck's neck and gently press upwards and downwards on each vertebrae. You will hear or feel a 'click' as the offset joint is placed back in position.

Massage the muscles around the neck to get the blood circulating and stimulate the area. Avoid the problem by using a suitable crate for transporting your birds.

Dry bill

Sometimes the area where the bill meets with the head of a duck will look very dry and cracked. Rub a small amount of butter on the affected area. This will moisturise the bill and have it looking brand new in no time.

Egg bound

Egg bound ducks are not uncommon in the heavy breeds, and typically the affected duck will be your best layer. The problem can be caused by an unusually large or misshapen egg, or when weakness or overworking by a drake during mating season results in the duck not being strong enough to discharge the egg down through the oviduct. You will know when a duck is egg bound because she will be straining and pressing, apparently uncomfortable and distressed, and constantly visiting the nest. The abdomen will have dropped dramatically.

Treatment: quietly catch your duck and run your fingers under her breast and towards the vent. About 7 cm in front of the vent opening, gently push to feel for the egg. When you have located it, use Vaseline to grease the vent or protruding organs (if prolapse has occurred). Sometimes you can carefully and gently massage the egg along the oviduct and out through the vent, by gently exerting a backward pressure until one of the poles of the egg appears. Slide the egg out, do not force it.

It is a good idea to tube administer approximately 10 mL of molasses, cod-liver oil and ground coral lime or limestone for a few days afterwards.

Foot problems

The underside of the web of a duck is a lot softer than the feet of a land bird. Consequently, some ducks get sore feet from running on hard ground and dry surfaces that bruise their pads. However, the most common cause, although not always recognised, is a dietary deficiency in biotin, pantothenic acid or one of the other vitamins and minerals important to the maintenance of healthy skin.

Often large corns and rough calluses develop on the underside of feet. In severe cases, deep bleeding cracks are evident and the pads swell and become infected, causing the bird to go lame.

Leg and hip bones of birds that continually run on hard ground can also be damaged by the jolting to which they are subjected. A comparison would be us walking on a hard, cold concrete floor all day without the correct footwear to cushion our stride. The end result would be hip problems, cracked soles and probably varicose veins.

Treatment: if dietary deficiencies are suspected, supplement the birds' diet with rations of vitamin premix or foodstuffs. Brewer's yeast, dried skim milk or lucerne alfalfa meal are all high in vitamin A and biotin. If the foot is inflamed, open the callus with a sharp, sterilised scalpel, remove the pus or core, and disinfect with alcohol or hydrogen peroxide. Place the duck in a clean pen with a thick layer of clean straw. Provide a well-balanced diet of green feed and a small container of water so the bird cannot make a huge mess. Wash and disinfect the wound each day. An application of Stockholm tar or moisturiser may help promote fast healing.

Prevention: a duck on a balanced diet with soft grass to walk on will not have foot problems.

Male prolapse

Sometimes a drake's penis (pencil) fails to retract. This can be common in older drakes. Carefully isolate the drake, and wash his penis with warm water and disinfectant, then place lots of antiseptic cream on his penis and gently replace it back into the abdominal cavity. If it is swollen, ice cubes can be used to treat the swelling. Ensure that there is a clean, cool pond for him to swim in during the next week, and keep a watchful eye on his progress. If not treated, the drake will eventually die of blood poisoning, as his genitalia will rot. I have heard cases of the penis dropping off and the drake surviving, although obviously he will be unable to service any ducks. A teaspoon of caster oil down the throat is advisable, along with light meals for a week. No further treatment should be necessary.

Medication poisoning

Ducks are easily poisoned when fed medicated feed designed for fowls or turkeys. Their systems absorb things they ingest more readily than the systems of land birds, and because they shovel their food and eat more than fowls, they readily overdose. This is worth keeping in mind when the ground is sprayed for parasites. If ducks are in the vicinity, they too will be killed. Affected birds will lose their appetites, become weak, have stunted growth, regurgitate their food and die suddenly.

Treatment: at first sign of these symptoms, switch to non-medicated feed, provide an abundant supply of succulent greens and mix rolled whole grains with the feed.

Rubbish and wire

Ducks will swallow nails and bits of wire, or pieces of string. When ingested, these objects often become impacted or puncture some part of the digestive tract. Affected ducks will slowly lose weight, stop eating and sit with their eyes partially closed, in severe pain. When a post-mortem is performed, the foreign matter will be found somewhere in the oesophagus or gizzard. But too late!

Treatment: is not possible unless the problem is in the crop, in which case the crop can be slit open and the object removed. If the bird has cut itself open on the body, you can stitch the wound closed using a sterile needle and some cotton thread. It will not be necessary to remove the cotton as the wound will heal cleanly and the cotton will fall out.

Slipped wing

Sometimes the wing tips flip outwards. There are many views and opinions as to why this happens. Feeding ducks turkey starter or any feed that is too high in protein is thought to be a possible cause of slipped wing. Some breeders believe that it occurs when pin-feathers on the young bird's wings grow too quickly, becoming heavy and turning the wing out. Others suggest it is the result of too much inbreeding, and still others believe it is due to ducklings grasping each others' wings and dislocating the wing tip.

My personal belief is that the underlying cause is not hereditary. I have had ducks and geese with this problem and none of them have reproduced offspring with slipped wings. I favour the view that it occurs when protein levels are too high. High protein promotes feather growth, and the flight feathers grow too heavy for the young wing muscle to support them, causing the wing to drop or flip out.

Treatment: is successful if noticed early and prompt action is taken. The problem can be corrected by folding the feathered limb in the correct position and taping it against the body, or by fitting the bird in a 'body stocking' created by cutting feet and neck holes in a stocking. With both these treatments the wing will become stiff from disuse, so the bandage or stocking should not be left in place for more than 4–5 days.

Soft-shelled and deformed eggs

Eggs that are laid with no shell or soft shells, eggs that break at the slightest touch, and those that have a long calcified tail on one pole are likely to be attributable to the duck's diet. The tail on the end of an egg is the shell from the following egg that has been attached to the one being laid. If not treated and corrected, ducks will continue to lay deformed eggs.

Treatment: place ground limestone or ground coral lime in a familiar container and leave it in the yard for the duck to eat. Alternatively, mix the limestone or coral lime into the duck's feed.

Spraddled legs

Spraddled legs are caused by running ducklings on slippery surfaces where they have difficulty gaining a footing. Under these conditions, the birds' legs will go out to the sides and not support their weight.

Treatment: use sticky tape to fix the legs into the correct position or tie the legs together with a short piece of string. When hobbles are used make sure you don't cut off the blood supply to the ducklings' legs. Keep the ducklings warm and comfortable, and ensure they are eating and drinking, Cover the bottom of the brooder with something firm and rough on which they can gain a footing. Treated ducklings usually right themselves within 2 days.

If a duckling hatches with splayed legs (the legs seem to grow out to the side of the bird), you can tape the legs in the correct position for a day or two. As ducklings grow so rapidly, if this problem is going to be corrected, it will be fixed within 2 days.

11
Duck egg incubation

Egg collection

Eggs that are to be incubated either artificially or under a duck other than the laying duck should be collected daily. This ensures they are fresh, protects them from crows and stops them becoming dirty. It also means your laying duck is less likely to become broody and sit on the eggs herself. This is especially true of Muscovies. They are natural breeders known for their mothering ability.

If some eggs are soiled, use a small piece of dry steel wool to clean them. Do not wash duck eggs. The shell is very porous and will absorb any detergents or disinfectants that are used. It is covered with a very thin cuticle-like substance that prevents bacteria entering, and seals in moisture for the developing duckling during its incubation period. If eggs are badly soiled, wash them in a solution of Condy's crystals to remove any bacteria that may be on the shell.

Egg selection

When selecting eggs to set, choose ones that are of a good size, well formed and smooth, and with no calcium build-up on the shell. If an egg is too small (such as an egg laid by a young duck), although it may be fertile, the developing duckling grows the same size as it would in a normal egg, then runs out of room before the incubation period is complete. This often results in death before hatching. Extremely large eggs may have double yolks and should not be set.

On a number of occasions, I have successfully placed small pieces of sticky tape over small accidental cracks when incubating valuable eggs. Just be sure to keep a watchful eye on the progress of any egg patched in this way. If bacteria infiltrates the shell, the egg will go rotten very quickly and smell, or may explode in your incubator ... and that is not nice!

Storing eggs for setting

Fertile eggs can be stored for up to 14 days. The ideal temperature for storing them is 12.8–18.3°C, so keep them in a cool, dark place, but not in the refrigerator. Low temperatures will cause the eggs to die; higher temperatures can commence incubation.

If you intend to keep the eggs in storage for more than 14 days, lower the temperature to 3°C.

When a duck commences her broodiness, she will lay one egg a day for approximately 2 weeks. Each day as she enters her nest, the eggs already laid are disturbed when she moves them to lay another. It is recommended that you emulate this natural movement of the eggs by turning them at least once a day, as this prevents the yolks from setting in the shells, rendering the eggs useless.

There are two stages of incubation. The first is a setting period of 25 days for all breeds other than the Muscovy, which has a setting period of 32 days. Hatching occurs in the final 3 days.

Candling to test eggs for defects prior to setting

Duck eggs are very easy to candle by shining a light through the shell in a dark room. Once you have built a light box (see Figure 11.1), you are ready to candle your eggs.

Hold the egg to be checked over the pipe coming from the side of the light box. The light will pass directly through the shell, illuminating the inside of the egg.

Common defects include:

- **Stale eggs:** will have a larger air cell than usual at the blunt end of the egg, and the yolk will be denser with a uniform outline.

- **Meat or blood spots:** are red spots appearing to float on the side of the yolk. This has nothing to do with a developing embryo, but is the consequence of a little bleeding inside the duck while the egg is being produced, high up in her ovary.
- **Double yolks:** can be easily seen. These eggs should be discarded. They rarely hatch, there being insufficient room for two ducklings to fully develop in one shell. Also, most double yolks are infertile.
- **Watery whites:** look as though there are little bubbles in the whites, and these change position when the egg is moved around during candling. Some birds produce sequences of watery eggs.
- **Jupiter eggs:** have defined swirls through the shell that can be seen during candling. These eggs are rarely fertile and never hatch.

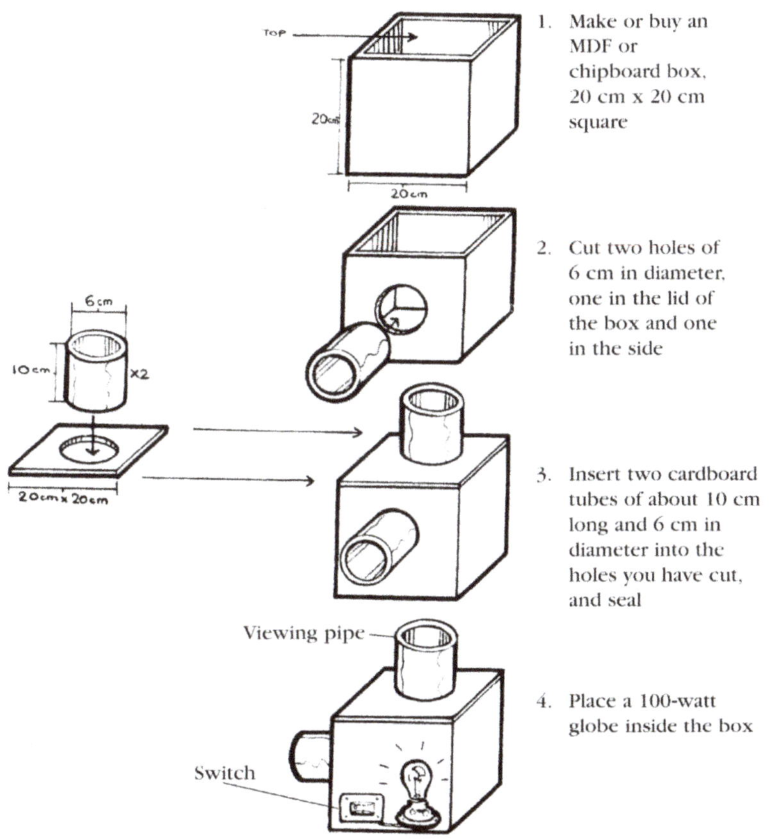

Figure 11.1: How to make a lightbox.

- **Excessive calcium:** causes irregular surfaces on shells. The shell of the egg should be smooth with no lumps, bumps or abnormalities.
- **Soft shells:** are evidence of a calcium deficiency and are obviously of no benefit. Rectify the production of soft shells by placing some limestone with the feed.

When a female duckling hatches, her system is already genetically programmed to place a specific amount of calcium around her eggs for the rest of her life; it may be a teaspoon or more. Nothing will change her genetic makeup, and if her dietary needs are not catered for, she will not be able to produce the required amount of calcium for her eggshells.

Once you have chosen some smooth-shelled eggs that are a good size, set them in your incubator.

Incubation

Natural incubation

Natural incubation is the most uncomplicated hatching of all and is discussed under 'Nesting behaviour' in Chapter 6.

Figure 11.2: Muscovies have a high degree of broodiness; they go broody regularly. They are also very good mothers.

Artificial incubation

Types of incubators

There are many and varied incubators on the market catering for eggs of different sizes and with various egg capacities. Basically, though, they boil down to two types: still-air incubators and fan-forced incubators.

Still-air incubators are electric and are usually made from foam boxes, although they can be made from just about anything. I have seen them in wood and steel. The heat source is an element suspended above the eggs with a capillary (heat sensor) hanging over the eggs to read and gauge the temperature and relay this to the thermostat, which controls the heat by turning an element on and off.

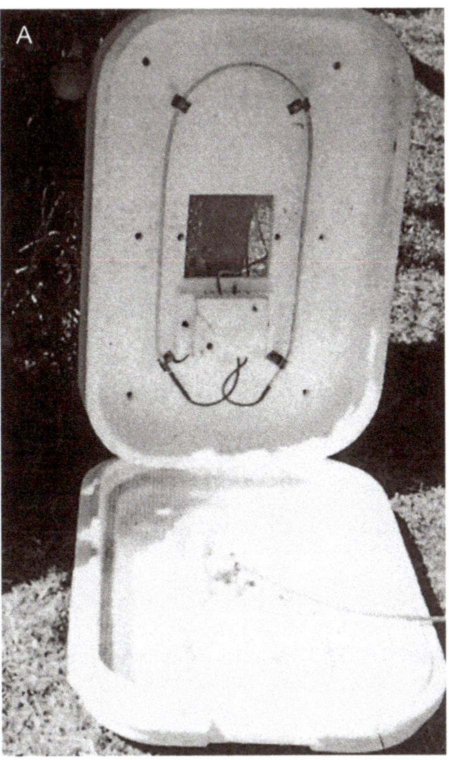

Figure 11.3: (A) The Bellsouth 100 still-air incubator has the capacity to hold 50 duck eggs. It costs around $300 to buy new. Cost to run is approximately 10 cents per day. (B) A Multiplo fan-forced incubator that will hold 50–60 duck eggs, depending on the size of the eggs. New, this incubator would cost around $1200 to buy. Cost to run is approximately $1.20 per day.

I use still-air incubators and have done so for some years, finding them easy to operate, dependable and almost maintenance free. They come in capacities that accommodate approximately 100 eggs.

The fan-forced models are fitted with fans that circulate heat and air. They have multiple shelves and trays with space for a couple of hundred eggs, which are turned automatically. Fan-forced incubators take up less room than still-air models when large quantities of eggs are involved, but both maintenance and purchase costs can be quite high.

Apart from off-the-shelf models, incubators can be homemade or purchased in kit form. Some poultry equipment businesses stock kits that can be fitted to wooden, steel or foam boxes. Bee boxes also make good incubators. Incubators used to be run on kerosene and were quite ingenious.

In an emergency, it is possible to hatch eggs in an electric frying pan, providing the temperature doesn't exceed 39.4°C, and I know a schoolgirl who hatched quail under a light globe ... no mean feat!

It is possible to have more than one setting of eggs, or hatch date, in your incubator at a time. I have at least one hatching every week, so I pencil the hatch date on the shells when placing eggs in the incubator.

Incubator ventilation

Ventilation and consistent temperatures are very important factors when hatching ducklings, so incubators are best kept in a well-ventilated room where the temperature doesn't fluctuate too much.

If, for example, you place an incubator in a laundry with poor ventilation and high humidity, the size of the air cell in the eggs decreases, the fluid level in the shell rises and the developing duckling drowns before hatching.

Incubator temperature

The ideal temperature for still-air incubators is 39.4°C, and 37°C for a fan-forced incubator.

I allow the eggs to warm up with the incubator, as opposed to warming the incubator and then adding the eggs. My reasoning is that I am simulating the natural process where a duck sits on a cold egg and warms it with her body heat. Keep a watchful eye on the incubator to ensure it reaches the required temperature.

When still-air incubators are used, it is recommended that you air, or cool, the eggs from time to time. You can do this by simply lifting the lid on your incubator and allowing cool air to circulate. I make a practice of using my candling time to air the incubator. While I test eggs under the light, I leave the lid of the incubator up for 5 to 10 minutes. I view this as similar to a duck leaving her nest to eat, drink and swim.

You must protect the eggs from temperature extremes during the hatching period, and over the final 3 days raise the humidity to 75%, sprinkling the eggs with water daily. Increase the airflow during hatching and leave the incubator closed.

Incubator humidity

Humidity is very important, especially in a country as dry as Australia. When you set your eggs, initially place a very small amount of water in the trays, dish, or channels provided in your incubator for this purpose. One centimetre of water in an incubator tray is sufficient for an average of 50 duck eggs per tray. The eggs can also be sprayed twice daily with a fine mist of lukewarm water. Many references state that this softens the shell for the duckling to pip through, and simulates the sprinkling the eggs would receive when the duck has swum then preens while standing over the eggs on returning to the nest.

My locality in Western Australia is quite humid in comparison with some parts of Australia, so it is only in the final 3 days that I add water to the trays and spray the eggs. The Bellsouth incubators I use are basically foam boxes with a thermostat and a heating element. I found that when I added water during incubation, my ducklings were often slimy at hatch, indicating a humidity problem. I improved the ventilation and lowered the humidity by enlarging the ventilation holes in the lid of the incubator using a half-inch bit. This solved the problem.

Turning the eggs

Eggs of all duck breeds (other than Muscovy) must be turned twice a day for the first 25 days of incubation (32 days for Muscovy eggs). It is not necessary to turn the eggs exactly 180 degrees, or to mark them with a pencil so you know which end was up before you turned them. Simply run the palm of your hand across the eggs, moving them gently. This prevents the embryo sticking to the shell, and moves toxins and by-products away from the embryo and nutrition towards it.

Leave the eggs alone in the final 3 days before hatching. At this stage the ducklings are positioning themselves to hatch. This is known as the hatching period.

Candling the eggs during incubation

Candle the eggs to test for fertility and determine which eggs have not taken. Fertile eggs will have what appears to be a contact lens floating in the yolk. With experience, the fertility of eggs can be checked within 24 hours

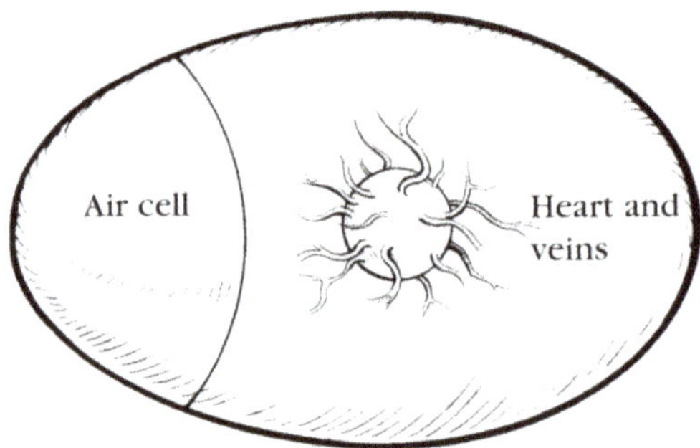

Figure 11.4: A healthy egg at 14 days.

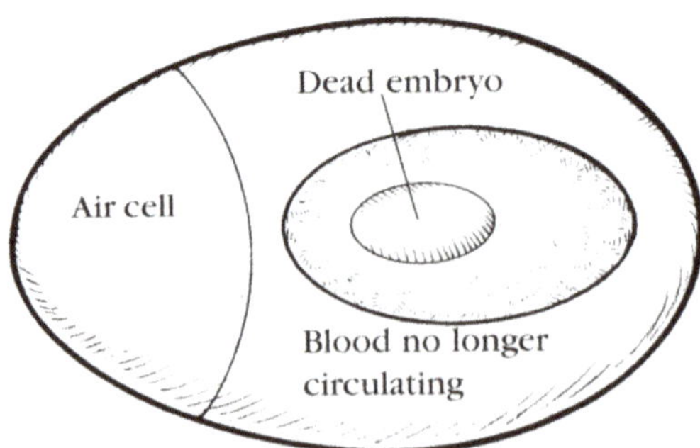

Figure 11.5: This egg has been incubated for 4 days. Obviously the circle of blood is not nourishing the heart. Development has terminated and the egg will go rotten. Throw it out.

of laying. Beginners will find that 4 days after laying is the ideal time to candle eggs, as by that time a small black or red dot can be easily detected. This is the heart, already formed and beating, with a net of blood vessels surrounding and extending away from it. These signs indicate a healthy developing embryo.

The air cell is where the duckling should pip through and hatch; its size is determined by the humidity in the incubator. I say 'should', as sometimes the developing duckling rolls to the pointy end of the egg. This is often caused by excessive humidity.

Eggs in which development has ceased should be disposed of; also immediately discard any eggs with a black dot and a blue blob inside. This indicates a fungal growth and it will spread to the good eggs, killing the embryos. If you find an egg with this fungus, the incubator should be cleaned and thoroughly disinfected.

It is advisable to candle the eggs every 4 or 5 days to ensure that the embryos are developing normally. By the 25th day you should be able to see a healthy air cell at the blunt end of the egg. The membrane that meets it should have a strong vein of blood running down to the duckling's body, which will be a large black area in the egg. Sometimes you will observe the duckling's bill rocking back and forth towards the air cell.

Incubation problems

- **Blood rings seen in eggs when candling**
  Causes: faulty storage of eggs or irregular incubation temperature.
- **15% of eggs infertile**
  Causes: can be due to too few drakes, old or crippled breeders, immature birds, breeders constantly being disturbed, no swimming water, first eggs of the season, medication in feed or water and/or the eggs having been stored for more that 14 days.
- **Ruptured air cell**
  Causes: rough handling or a deformity. Always handle eggs gently.
- **Yolk stuck to side of shell**
  Cause: old eggs that have not been turned regularly.
- **Dark blotches on shell**
  Cause: contamination from bacteria on shell at the time of collection.

- **5% or more dead embryos between 7th and 25th day**
  Causes: poor breeder diet, inbred stock, faulty temperature in incubator and/or inadequate turning of eggs.
- **Early hatching**
  Cause: high incubation temperature.
- **Late hatching**
  Cause: low incubation temperature.
- **Eggs pip, but don't hatch**
  Causes: high or low humidity affecting the egg membrane, eggs chilled then overheated during last 3 days of hatching, poor ventilation and/or disturbances during the hatching period.
- **Eggs pip at the small end**
  Causes: eggs incubated in the wrong position (position them with the small pointy end lower than the large blunt one), air cell up due to too much humidity and/or poor ventilation.
- **Sticky or slimy ducklings**
  Cause: high humidity during incubation and hatch. The slime is the white membrane that hasn't been able to dry out.
- **Large or protruding navels**
  Causes: high temperature, excessive dehydration of eggs and/or bacterial infection. Lower the temperature and check thermometers. Increase the humidity level and improve sanitation.
- **Dead ducklings in incubator**
  Causes: suffocation and overheating. Increase the ventilation during the hatch and watch the temperature carefully.
- **Spraddled legs**
  Causes: smooth incubator trays or brooding floor and/or high temperature in the incubator. Cover trays or flooring with cloth or sawdust.
- **5% cripples**
  Causes: inadequate turning during incubation, prolonged periods of cooling and/or inherited defects.

If in doubt, throw it out

If your incubator smells foul, you have a rotten egg. When light testing your eggs, it is good practice to throw out any you are unsure of. My motto when candling eggs is: 'When in doubt throw it out'.

Rotting eggs can be costly in more ways than one. Often space in the incubator is at a premium, and an infertile egg is taking up valuable room.

Apart from exploding and/or spreading bacteria to otherwise good eggs, rotting eggs can also absorb heat, giving incorrect thermometer readings.

Fumigating your incubator

It is essential that the incubator is clean when you set eggs. It is possible to fumigate an incubator with eggs still in it using Condy's crystals and formalin in accordance with the manufacturer's instructions. Or you can follow my practice of emptying the incubator after each hatch and scrubbing it out with bleach and a cream cleanser such as Ajax or Bon Ami. The cleanser will scour off any yolk and blood that may have collected at the bottom of the incubator, and the bleach kills any bacteria floating around. I also scrub the incubator thoroughly if any spillage of blood or ruptured yolk occurs.

From the moment it hatches, a duckling begins to moult very fine down feathers. These collect in the water at the bottom of the incubator and bacteria begin to grow on them. These bacteria can harm your eggs. They penetrate the shells and destroy the embryos.

I know you are probably thinking that when a duck sits on her eggs she is not in a sterile environment and her eggs hatch without all this fuss. This is quite true, but eggs incubated by a duck have a great deal more ventilation, the nesting material is moved around constantly, and the duck monitors her eggs very closely, rolling out any that are infertile or dead. A duck keeps her nest in order, whereas a human incubating duck eggs sets up a warm and humid environment that is ideal for encouraging harmful bacteria to flourish.

Experiment with your incubator

Don't be frightened to experiment with your incubator. I incubated duck eggs at 35°C and found that some ducklings hatched in 30 days – 2 days longer than usual – while some never hatched at all. With the next batch, I increased the temperature to 41°C. The ducklings commenced hatching 4 days early and all survived.

I am not suggesting that you raise or lower the temperature of your incubator from the optimum, but if you strike problems with your ducklings hatching, ask yourself what has changed. Have you moved your incubator? Are you constantly lifting the lid and allowing the temperature to fall and humidity to drop? Has a power failure caused the temperature to drop? Is the temperature consistent? Is the incubator in a well-ventilated room? Is the thermometer registering the correct temperature? Did you have a bad egg in the incubator?

12 Hatching and rearing ducklings

Hatching ducklings

To keep the incubator clean, place hatching eggs in a small shallow-sided hatching tray in the incubator with other eggs, or in another incubator on their own. When they hatch, the seemingly tonnes of down they shed will not be all over the bottom of the incubator. To clean up, simply remove the tray.

Ducklings that have no trouble hatching should be removed and placed in a brooder to dry. Remove their empty shells so these cannot be kicked around and pushed over another egg, preventing it from hatching.

When you open the incubator to remove hatched ducklings, the humidity level in the incubator falls. Correct this by placing about half a cup of almost boiling water into the incubator water trays. The resulting steam will replace the lost humidity.

It is not advisable to assist a duckling to hatch; however, we all do it if we see one in difficulty. A friend of mine is a whiz at successfully hatching problem ducklings. Before assisting, she determines if the duckling is making excessive and distressed noises inside the shell. If you decide to help a duckling out, assess the situation first by taking the following steps.

Light test

Hold the egg up to your lightbox and look for the duckling's bill to see if it is rocking back and forth. If so, carefully break the shell exactly where the bill is. This will give the duckling a breath of air.

Examine the membrane encasing the duckling

If the membrane is white and fluffy and you can clearly see the veins of blood, carefully and gently chip the shell only in a circle around the air cell of the egg, without breaking the white membrane beneath it. Assess the amount of blood in the membrane. If it bleeds when you make a small tear in it, apply castor sugar to the bleeding area, as this will clot the blood. Place the egg back in the incubator and wait an hour or so.

If the membrane is yellowish and drying, carefully make the hole in the shell larger and peel back the membrane towards the shell, like peeling a banana. Do not tear the membrane.

When you return, examine the colour of the duckling's skin beneath the feathers. If it is red, the duckling can be removed from the shell. When removing a duckling from its shell, look down its breast to the umbilical cord, and make absolutely sure the yolk has been absorbed into the young bird. If the yolk is still visible, leave the duckling in the shell. The yolk will not be absorbed once the bird has been removed, and eventually the duckling will die.

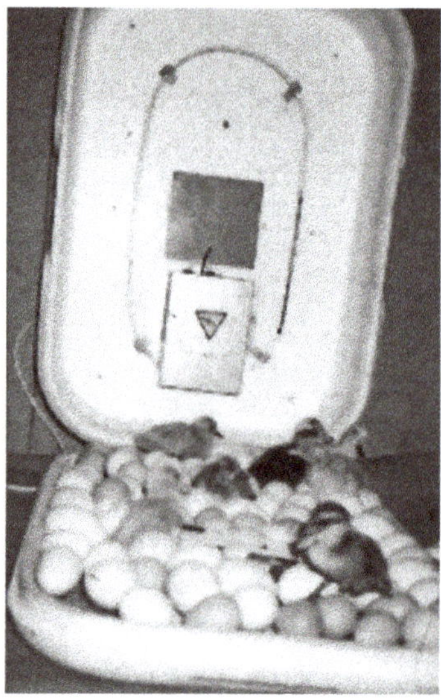

Figure 12.1: Ducklings hatching in a Bellsouth 100 still-air incubator.

Once the duckling has been removed from its shell

If the duckling is weak, fill a basin with some water that is at body temperature, and while holding the duckling in the palm of your hand, swim it through the water. Try to make it paddle a little and wash any blood and rubbish from its feathers. It may even take a small amount to drink. If it begins to cheep, this is a good sign.

Caster sugar can be used to halt bleeding by placing a small amount on the umbilical cord. You can also mix a very small amount of caster sugar in water and entice the duckling to drink a little. This will give it glucose, which will stimulate the duckling almost immediately. Return the duckling to the incubator for the warmth it requires until it has fluffed up and dried out.

Sometimes a duckling will hatch with its head swaying or throwing backwards. This can indicate a low sugar level to its brain. Perseverance will prevail if you continually give it small sips of sweetened water. Sometimes this treatment might need to be maintained for 3–4 days before the duckling improves.

Imprinting at hatching

When artificially hatching eggs, it is possible to imprint your voice on the developing ducklings by repeating a phrase each day when you turn the eggs. I have found by repeating 'little ducks' to the eggs, the hatched ducklings recognise my voice and come running to me when I call 'little ducks'. (It's really quite cute and touching and I honestly don't think there is anything more appealing than a new little duckling.) This suggests that ducklings grow accustomed to their mother's sounds during her sitting period, so even when newly hatched they have no trouble recognising her.

Raising ducklings

Raising ducklings in a brooder

You can make your own brooder by placing a light globe in a box. The light globe will generate enough warmth to keep the inside of the box at a comfortable temperature, including at night. The ideal brooder temperature is 32°C; however, I simply hang a light globe about 30 cm from the ducklings, and if they are cold, they move beneath it, if they are hot they can move away. We have found that a white light in the brooder box is more beneficial than an infra-red one. After a week under a 75-watt infra-red light, our ducklings became a little disorientated.

Figure 12.2: Appleyard and Pekin ducklings, owned by Nyiri Murtagh.

Figure 12.3: A 100-watt incandescent light globe is used to brood the ducklings in this brooder. Pine shavings are used as litter.

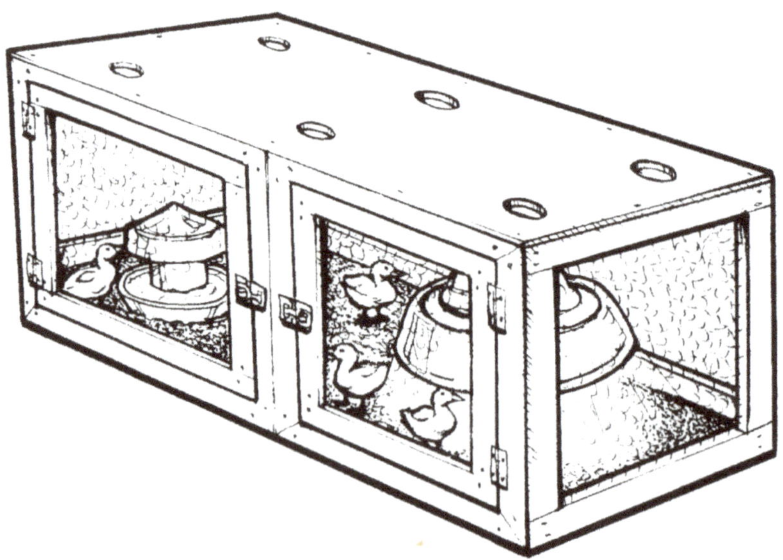

Figure 12.4: A simple brooder box can be easily made at home. A brooder 1 m × 0.6 m would comfortably house 10 small ducklings.

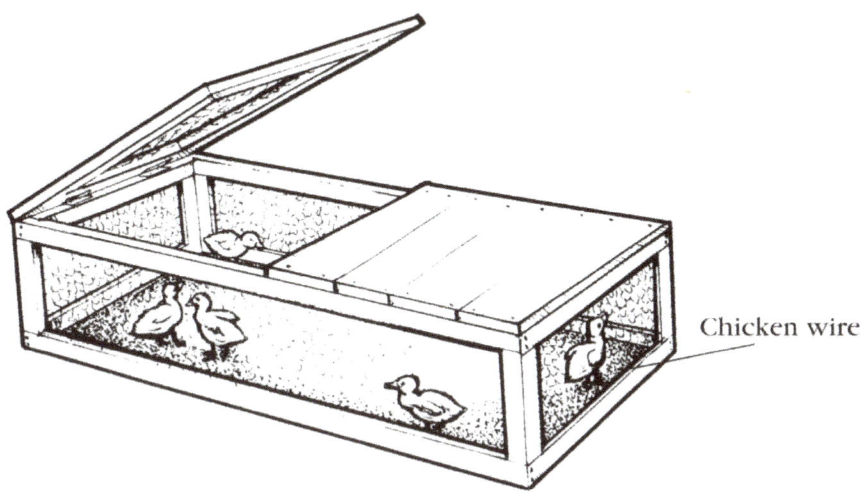

Figure 12.5: A dry, draught-free area is all your ducklings need after about 10 days.

Ventilation is a must in the brooder box as ducklings often make a mess with their drinking water (which I should not leave in overnight, but I do). In the morning when I greet them, I wonder how they can make a little water go so far. If ventilation is inadequate, the humidity and warmth that builds up in the brooder poses a health risk to the ducklings. The same applies for goslings and chickens.

The litter on the brooder floor should be sawdust. It's absorbent and doesn't mat or become smelly like straw does. Never run ducklings on newspaper or any slippery surface, or you will have ducklings with splayed legs.

Waterbirds run a very high body temperature, and unlike fowls, they retain and generate body heat. A duck's body temperature is 42.8°, and it is most unusual for a duckling to get sick.

After 10 days in the brooder, it will not be necessary to run a light for ducklings, providing there is more than one of them, and the area they are in is dry and draught free.

A safe cage can be made from wood and chicken wire. The cage shown in Figure 12.5 is bottomless to enable ducklings to pick at grass. The lid opens upwards and one side of the top is covered to provide shade. These runs can be moved around a lawn quite easily.

Figure 12.6: A duckling feeder (which has small holes and/or mesh to prevent wastage) and a waterer that will not allow the ducklings to get into it.

Make sure the waterers you use are not deep, and if you provide a bowl for ducklings to swim in, make sure their feet can touch the bottom, and that they can get out very easily without becoming waterlogged and risking drowning. I can't stress this enough. Far too often people have purchased young ducklings from us, only to take them away and have them drown.

Raising ducklings is really a matter of using common sense. For the first couple of weeks they only have down covering their bodies, and they are certainly not waterproof.

Do not run chickens with ducklings ... not because they disagree with one another, but because ducklings seem to have a saying, 'Everyone into the pool', and this will include your little chickens. When you open up your brooder the chickens will be wet, shivering little mops.

If the day is sunny and warm, you can run your ducklings outside, but watch out for cats and crows while the ducklings are still small. Be sure that your dog is under control as well; many dogs seem to think that ducklings taste good.

No water should be left in the brooder at night. This is viewed by the ducklings as a swimming hole for their own entertainment, and they will keep you busy cleaning up wet walls and sodden sawdust.

If you place ducklings outside, put them in an enclosure with a roof. The roof can be a wire one, or simply some shade cloth thrown over a little yard. Once a duckling is 3 weeks old it can sleep outside in a dry and draught-free shed or little house, whichever you provide.

Duckling diet

When ducklings hatch, they have absorbed enough yolk into their bodies to last them 2 or 3 days. But remember, a duckling is only as good as the feed you gave the duck that laid the egg. Although ducklings don't require food and water immediately, I like to get them eating and drinking as soon as possible.

Ducks are foragers and rarely stop eating, which isn't surprising when you consider how quickly they grow. Most breeds have attained at least half their adult size by the time they are 6–8 weeks old. Hence the importance of beginning on the right diet.

During the first few weeks it is advisable to allow the ducklings access to food whenever they feel hungry. I feed our ducklings broiler starter, which can be purchased from any produce store. Don't feed ducklings pollard or bran, but the vitamins and minerals they obtain from greens are a must for their growing bodies. Cabbage is a good vegetable to feed, being particularly high in iron.

Don't feed waterbirds turkey starter; the results can be disastrous. It is too high in protein (22%) and can contain a steroid or anti-coccidiostat. I have found waterbirds fed on it developed slipped wing (see 'Slipped wing' in Chapter 10).

At 4 weeks of age, the duckling should be in your poultry or duck yard. If grass and greens are not readily available, a capful of cod-liver oil can be placed in their feed. This is a source of B-group vitamins and vitamins E, A and K.

Three meals a day are sufficient for ducklings from 8–12 weeks old, although, as I stated earlier, I always leave food out for the ducks. However, if your birds range free, one meal a day could be sufficient. You will know if your birds are hungry by the end of the day, as a noisy duck is usually a hungry duck. It is a good idea to place a little feed in their house at night. They will home in on it very quickly and after a little time will automatically go into their shed each night.

Duckling-related problems and their solutions

Feather eating

Feather eating can be prevalent among ducklings, especially those that are brooded artificially. It can be the result of high brooding temperatures, excessive light, over-crowding, an unbalanced diet, insufficient quantities of feed or lack of roughage. Obviously the symptoms are that they pull out each other's feathers and eat them.

To solve the problem check the brooder temperature, reduce the intensity of light (using blue bulbs can help), and provide the ducklings with more space and a balanced diet with sufficient feed, and some greens.

Waterlogged ducklings

We are all subjected to that frightening moment when we discover a duckling half-drowned and suffering from hypothermia. The best thing to do is fill a basin with warm water and place the duckling in it, supporting it in your hand. You should see some life come back into its little body as its temperature comes up. Dry the bird off with a towel or a hair dryer and give it a small drink of warm, sugary water. Place the duckling into the brooder in a nice warm spot, or even back into the incubator for a while until it recovers.

Ant and spider bites

The only other major threats to young waterbirds are ants and spiders. As yet, I know of no serum to reverse the effects of these bites; however, it has been suggested that an antihistamine tablet can be broken into quarters and administered to an affected bird. You will know that a duckling has been bitten if it suffers fits, throwing its head backwards and stretching its legs behind itself.

13
Crossbreeding ducks

With the exception of the Muscovy, all the duck breeds mentioned in this book are descendants of the wild Mallard. Each breed is the result of careful selective breeding of Mallard sports that were thrown at various times. For example, the Aylesbury is thought to have originated as a white mutation of the Mallard, and was subsequently bred for increased weight as a table bird.

Developing any new breed of waterfowl is a long and unenviable task. For a breed to be recognised, it must be shown to have reproduced itself many times without any deviations or 'throwbacks' to the original parent stock.

Some breeds have been around longer than others; consequently, their characteristics are more fixed. Breeds such as the Elizabeth, Blue Swedish and Welsh Harlequin are relatively new on the scene and, despite being standardised, are considered little better than crossbreds by purist breeders. I confess that this is a view I hold myself.

You will sometimes hear breeders make reference to Mendel's laws, especially when one of their purebreds produces a mutation or a 'throwback' to a type found in its ancestry (one of the breeds used in the development of a breed). For example, sometimes a Buff Orpington will reproduce a 'blue' Orpington.

Gregor Mendel was an Austrian monk. Much of the modern science of genetics is based on what he discovered while crossbreeding strains of peas in an Augustine Monastery between 1856 and 1863. Mendel proposed that characteristics such as colour and size were due to the occurrence of paired units of heredity, now known as genes. He believed that certain features of colour, size, etc. remain dominant, while others are recessive. As only one-half of a parent's factors are transmitted to each offspring, different offspring from the same parents receive different combinations of hereditary factors.

While there is variation in all poultry, more so when domesticated, there is still a baseline or foundation. Broadly speaking, the offspring will resemble the parents in conformation, type, colour and character. For example, a breeder with Rouens will breed Rouens. If they keep Indian Runners, the offspring will be Indian Runners. Any variations will be minor, and will be culled if not up to show standards.

However, amidst some exhibition waterfowl, the points that seem to be so desirable in the show arena can compromise and even sacrifice the virility of the breed. Continued inbreeding in a bid to emphasise artificial points will exhaust a line, rendering it infertile and feeble.

While inbreeding does not of itself cause disease, it leads to refinement of the breed's constitution, reducing the stamina of the bird so that it is less hardy. I acknowledge that a certain amount of inbreeding is necessary to perfect a breed, but serious deformities can arise where there is a failure to introduce new breeders. For instance, dwarfism is a common result of inbreeding Muscovies. This can be prevented by 'outcrossing', which simply means introducing a new, not necessarily unrelated bloodline, or a completely different breed with similar characteristics.

Fresh blood strengthens dominant features, and when introduced to a flock from time to time will enhance and or maintain the size and vigour of a line.

Now and then, variations appear even in the purest of duck stock. Often brown or red feathers will be noticeable. This is a reversion to the red-brown feathers of the original waterfowl. If these variants were permitted to breed without humans selecting the breeding stock, in very few generations flocks would revert to their ancestral form.

It is accepted that drakes influence the external characteristics of the offspring, while the duck influences the internal qualities, such as temperament and body size and shape. This is taken into account when breeding for particular effects. Let's say you have Buff Orpingtons with a black or dark bill and you want to restore the yellow bill colour in your Buff Orpington line. To do so you would mate a Pekin drake (yellow bill) with your Buff Orpington duck. A very large percentage of the resulting ducklings will have brilliant yellow bills, along with hybrid vigour and a few other changes. However, if you wanted to produce a line of birds for the table, all the ducks would be Aylesbury; the largest white ducks.

Crossbreeding for meat

If you'd like to create your own line of table birds, you can experiment with different crosses. Obviously, what you'll be aiming for is a heavy bird that grows rapidly (thereby reducing the amount of time you have to feed it).

From a commercial processor's point of view, only white ducks are acceptable, as black feathers leave unsightly blemishes on the skin. This might be an important consideration for you, too.

You must be able to produce a fast-growing duckling, while maintaining high egg production in the parent stock. I believe that the best cross is a drake meat line with a dual-purpose female line. This will produce hybrid vigour.

In the early days of the table-duck industry, the Aylesbury-Pekin crosses were the most viable. The Aylesbury drake was used as the meat line with the Pekin duck. The result of these Aylesbury-Pekin crosses were called Australian Pekins. An Australian Pekin is distinguishable from a purebred Pekin by the yellowish tint in its bill and its lack of keel.

Australian Pekins exhibit good hatchability and low carcass fat, and, because the Aylesbury has softer feathers and whiter meat than the Pekin, they are easier to pluck. They are marketed from 6–9 weeks of age, when they are termed 'green ducklings'. These birds are managed so they grow large in the shortest possible time, so they can be processed before they develop pin-feathers on the tips of their wings. Pin-feathers are the most difficult feathers to remove. After 9 weeks, any ducklings held back from processing at the green duckling stage are not processed until the pin-feathers are fully formed. On average, green ducklings weigh between 2 and 2.3 kg, having converted food to meat at about a 3 to 1 ratio. The success of the Aylesbury-Pekin crossing led to the purebred Aylesbury almost disappearing in Australia.

Australian Pekin drakes were later crossed with White Indian Runners and White Campbell ducks to increase egg production, although this lowered the weight of the ducklings produced. Remember the golden rule: it is the duck that determines body size.

While other meat breeds can achieve a heavier body weight than the Pekin, none can match the growth rate, and when one stops to consider the cost of feeding the ducklings, the viability in terms of cost must be taken into account.

You could also try crossbreeding with Muscovies. The Muscovy is noted for its flavoured, pinkish meat. A Muscovy drake with a Pekin female will result in good egg production, and the ducklings will be heavier than purebred Pekin ducklings. You will, however, come up against a few problems. For a start, Muscovy drakes don't, as a rule, mate naturally with ducks of other breeds. Most meat producers 'milk' the semen of Muscovy drakes and artificially inseminate the Pekin ducks. And, because Muscovies are not descended from the Mallard, crossing them with other duck breeds produces sterile offspring. Alternatively, you could try putting a Pekin drake with a Muscovy female. Pekin, Orpington and Indian Runner drakes will produce high egg fertility in Muscovy eggs.

Figure 13.1: Aylesbury-Pekin crosses, bred by Nyiri Murtagh.

Rouen and Aylesbury drakes crossed with Muscovy ducks will produce eggs of low fertility. I say this having observed the matings of these breeds, and having proved it by running Rouens and Muscovies together during mating season. Not once in 10 years have they crossbred naturally.

The University of Western Australia has experimented with growth rates of different crossbreeds. The results are shown in Appendix 1.

Developing your own lines

If you're interested in crossbreeding ducks, don't hesitate to experiment. You never know, you might come up with the makings of a wonderful new breed.

In the early stages of my interest in ducks, I unknowingly bought a pair of Buff Orpingtons with a lineage that turned out to be less than pure. Since the birds were not purebred in the first place, I decided to experiment with them.

Creating 'Buff Orpingtons'

Since my not-so-pure Orpingtons had black bills, legs and webs, the first assignment I set myself was to change the bill, leg and web colour from black to orange. To do this, I mated a Pekin drake (Pekins have orange bills) with my Orpington duck. All the ducklings resulting from this mating resembled the mother in carriage and body type, but had orange bills and legs. Unfortunately, none of the offspring exhibited a good strong buff colouring; some had patches of buff on white, while others were buff with white primary feathers. However, I kept two young drakes and two young ducks that only had white primary feathers.

The following year, I mated the drakes with two ducks that were Mallard-Pekin crosses and I put a Black Indian Runner drake with the two young ducks. The drakes with the Mallard-Pekin ducks produced offspring with Dark Khaki Campbell colouring, lots of lacing and orange feet. The bills were green in the drakes and orange in the ducks. I let this line go. However, the ducklings from the pairing of the Indian Runner drake with the young ducks were of greatly improved colour. Their buff colour was stronger and deeper, and while their bills, legs and webs were orange, they retained the body type and shape of their mothers. The females were coloured particularly well, but while the young drakes were a nice seal-brown around the head, the colour did not form a tidy circle around the neck, but extended down the chest.

I kept a breeding trio of the Indian Runner line and bred them back to one another. The result was exceptionally good; they looked like purebred Buff Orpingtons. The carriage was slightly elevated at the shoulders with a nice depth through the stern of the bird. The ducks' abdomens almost touched the ground. I took these birds to a poultry 'show and tell' day. A waterfowl judge assessed the birds, and seemed to think they were exceptionally good in type and colour. He inquired as to where I had obtained the line.

Creating 'Appleyards'

I have bred 'Appleyards' purely by chance, having hatched a Mallard egg we found in our yard in a pool of water (as mentioned on page 89). The duckling that hatched grew up and crossed with one of my Pekins. When the resulting young crossbreds grew and began to colour up, all who saw them commented on how attractive they were and how nice they looked in the garden, so I called them 'Garden' ducks.

Figure 13.2: Most of the offspring produced by crossing a Pekin drake with a Cayuga duck will be black with a white bib.

Later, I realised that these Mallard-Pekin crosses looked exactly like Silver Appleyards. I've since found that some Pekin-Rouen crosses also result in birds that look exactly like Silver Appleyards. This experience has made me wonder about the purity of some of the Appleyards I've seen.

Creating 'Blue Swedish'

If you want to create your own 'Blue Swedish', cross a Pekin drake with a Cayuga duck. The majority of the offspring will be black with white bibs, but a percentage will be solid blue, and a small number will exhibit plumage with pencilled black rims and a white bib. By culling all the black progeny and breeding the solid blues and white-bibbed blues back to one another the following season, you would begin to see some exhibition Blue Swedish among the offspring.

Appleyards crossed with Rouens, then crossed the following year with Black Indian Runners can also result in 'Blue Swedish'. However, this is a long, drawn-out process; the instances of good Blue Swedish are rare and the progeny don't breed true.

Appendix 1. Breed selection table

Breed	Class	Eye colour	Bill colour	Leg/foot colour
Appleyard (Silver)	Heavy	Dark hazel	Yellow	Light orange
Aylesbury	Heavy	Brown	Pink-white/flesh colour	Bright orange
Black East Indian	Bantam	Black	Black with blotches of olive	Black
Blue Swedish	Heavy	Brown	Drake: green with a black bean Duck: blue	Brownish orange
Campbell (Khaki)	Light	Brown	Drake: green or bluish-green Duck: dark brown	Drake: orange Duck: body colour
Cayuga	Heavy	Black	Slate-black with a dense black saddle in the centre	Dull orange-brown
Crested Duck	Light	Various	Various White Crested: orange	Various colours White Crested has orange legs
Duclair	Heavy	Brown	Drake: green Duck: orange	Burnt orange
Elizabeth	Bantam	Brown	Drake: green Duck: yellow	Browny-orange
Indian Runner	Light	Brown or blue	Various White: orange Fawn: yellowish-brown	Various White: orange webs and feet Fawn: yellowish-brown webs and feet
Mallard	Small	Brown	Drake: bright yellow-green Duck: orange with black central patch	Brownish orange
Muscovy	Heavy	Brown or blue	Yellow or black	Orange
Orpington (Buff)	Heavy	Brown	Orange	Bright orange
Pekin	Heavy	Leaden blue	Bright orange	Reddish-orange
Rouen	Heavy	Dark brown/ hazel	Drake: bright green-yellow with a black bean Duck: dull orange	Burnt orange
Saxony	Heavy	Brown	Yellow	Dark yellow
Welsh Harlequin	Light	Dark brown	Drake: greenish-blue Duck: dark greenish-blue	Drake: orange Duck: orange-brown

Wing bar colour	Plumage colour/description
Blue	Drake: resembles a Mallard in colour, but brighter. It has a beetle-green head, neck and back, silver-white neck ring, deep claret chest and shoulders, and blue wing bars. Duck: silvery-white, with a heavy flecking of fawn on her back.
Nil	Both sexes are snow-white.
Nil	Both sexes are black with a bottle-green sheen. Drake exhibits especially strong green sheen on head, neck and back.
Nil	Both sexes have blue bodies, white bibs and white outer-wing primaries.
Brown	Drake: body even khaki; head, neck, stern and wing bar lustrous green-bronze. Duck: warm khaki throughout body; neck and head a slightly darker shade.
Nil	Both sexes have deep black to green-black plumage.
Nil	Any colour is acceptable. White is popular.
Blue	Drake: identical to the Mallard in colour. It looks like the Rouen, but is lighter and has no keel, and is often confused with the Appleyard. Duck: chestnut with each feather pencilled in black.
Purple/blue	Drake: resembles a Mallard in colour, but brighter. Looks like a small Appleyard. Duck: silvery-white with chestnut flecking.
	Various colours including Fawn and White, Fawn, White, Brown or Black.
Purple/blue	Drake: head and neck bottle-green down to a white ring that encircles the neck; breast deep claret; stern and abdomen French grey-blue; flanks, sides and wing coverts clearly marked with black; back and rump a rich green-black; tail slate-black, purple and white wing bars. Duck: ground colour of chestnut-brown, with each feather pencilled in black; same wing bar colour as the drake.
Nil	A number of colour variations including White, Black and White, Blue, and Blue and White.
Nil	Deep reddish-buff. Drake has seal-brown head and neck, and red-brown rump.
Nil	Creamy white.
Drake: blue Duck: white	Drake: the same basic colour pattern as the Mallard drake, but is brighter in colour. Duck: the same colour pattern as the Mallard duck.
Blue	Drake: has a pigeon-blue head and neck, a white neck ring, a rusty red breast, oatmeal tail and wing feathers, and blue wing bars. Duck: a soft buff colour.
Green	Drake: head, neck, tail and rump black-green with bronze lustre; white neck ring; claret breast and shoulders; cream keel; green wing bar. Duck: fawn head and neck; red-brown to claret breast: white undercarriage; green flecks through to tail; green wing bar.

Appendix 2. Growth rates of Pekins and Muscovies

The University of Western Australia experimented with growth rates of different crossbreeds. The results are shown below.

Performance of Pekin

At 8 weeks of age the drakes weighed 2.28 kg, the ducks 2.15 kg.
The drakes consumed 6.51 kg, ducks 6.56 kg.

The drakes' cumulative conversion 2.9 kg, ducks' 3.11 kg.

At 14 weeks, the drakes weighed 2.82 kg, ducks 2.65 kg.
Drakes consumed 13.5 kg, ducks 14.39 kg.

The drakes' cumulative conversion 4.85 kg, ducks' 5.52 kg.

Performance of Muscovy

At 8 weeks of age the drakes weighed 2.28 kg, the ducks 1.58 kg.
The drakes consumed 5.81 kg, ducks 4.98 kg.

The drakes' cumulative conversion 2.6 kg, ducks' 3.24 kg.

At 14 weeks, the drakes weighed 3.99 kg, ducks 2.21 kg.
Drakes consumed 15.64 kg, ducks 9.9 kg.

The drakes' cumulative conversion 3.96 kg, ducks' 4.57 kg.

Performance of Muscovy male with Pekin female

At 8 weeks of age, the drakes weighed 2.51 kg, ducks 2.41 kg. Figures available for mixed sex: consumed 6.51 kg.

The cumulative conversion 2.5 kg.

At 14 weeks, the drakes weighed 3.51 kg, ducks 3.29 kg.
Drakes and ducks consumed 14.3 kg.

The cumulative conversion 4.26 kg.

Appendix 3. Determining sex by incubation temperature

Recently I carried out an experiment with eggs from different duck breeds: Pekin, Rouen and Indian Runner.

I set 100 eggs to hatch at varying dates. I marked the breed of duck on each egg, the date it was due to hatch, and how far into incubation the egg was. I divided the 100 eggs into four groups of 25:

- Group 1 eggs were 4 days into incubation.
- Group 2 eggs were 10 days into incubation.
- Group 3 eggs were 16 days into incubation.
- Group 4 eggs were 26 days into incubation.

I bought a thermometer for taking animal temperatures from the local veterinarian. It is digital and beeps when the temperature reading has been reached.

All the eggs were placed in an incubator reading 39.4°C. The eggs varied in size, shape and colour – as eggs do. However, I found that when I placed one egg at a time in an egg carton, air cell up, and slid the thermometer down the side of the egg, between the carton and the shell, regardless of the differences in incubation periods, each egg recorded a temperature of its own.

I found this somewhat incredible. It reminded me that I had heard that the crocodile lays her eggs in a circle, and the sex of the young is determined by whether they are towards the outside of the circle, where the temperature is cooler, or in the middle, where it is warmer. This supports evolutionary evidence that ducks are indeed related to reptiles, albeit distantly, and the more I study and observe my ducks, the more clearly I see the link.

Test results

ROUEN		INDIAN RUNNER		PEKIN	
Egg temperature (°C)	Sex of duckling	Egg temperature (°C)	Sex of duckling	Egg temperature (°C)	Sex of duckling
32.41	duck	33.82	duck	32.24	duck
33.08	duck	33.93	duck	32.25	duck
33.10	duck	34.46	duck	32.30	duck
33.21	duck	34.47	duck	32.33	duck
33.25	duck	34.49	duck	32.41	duck
33.30	duck	34.53	duck	32.43	duck
33.40	duck	34.56	duck	32.43	duck
33.47	duck	34.62	duck	32.43	duck
33.49	duck	34.68	duck	32.48	duck
33.51	drake	34.76	duck	32.53	duck
33.59	drake	34.76	duck	32.54	duck
33.59	drake	34.87	duck	32.63	duck
33.62	drake	35.03	drake	32.65	duck
33.77	drake	35.08	drake	32.65	duck
33.77	drake	35.11	duck	32.67	duck
33.82	drake	35.11	drake	32.72	duck
33.95	drake	35.15	drake	32.76	duck
34.29	drake	35.17	drake	32.86	duck
		35.22	drake	32.89	duck
		35.25	drake	33.11	drake
		35.30	drake	33.15	drake
		35.33	drake	33.62	drake
		35.37	drake	33.73	drake
		35.40	drake	33.76	drake
		35.52	drake		
		35.53	drake		
		35.68	drake		
		35.69	drake		
		35.70	drake		
		35.72	drake		
		35.78	drake		

You will notice that all the Rouen eggs lower than 33.5°C hatched as ducks, and all those above 33.5°C were drakes.

The Indian Runners ran a higher temperature again, even in the ducks. All ducks were under 35°C; the drakes recorded temperatures above 35°C. There were plenty of drakes.

All the Pekin eggs that were below 33°C hatched as ducks, those above were drakes.

I wondered whether there was any relevance in the differences of heat in the breeds. For example, the Indian Runners, on average, held a much higher temperature than the Rouens and Pekins. The Indian Runner, of course, is a light breed as opposed to the heavier Rouen. Regular observation of temperature has shown this premise to be correct each time, so it is no coincidence.

I suppose some breeders could use this to their advantage by limiting the number of drakes bred and gaining maximum space in the incubator by continuing with low-temperature eggs. However, it does take two to tango, and what if the perfect drake was among the drake eggs discarded?

Glossary

This glossary covers show terms as well as the terms used in this book.

abdomen: upper part of body, from keel to vent

AOC: any other colour

AOV: any other variety

axial feather: small feather between primary and secondary wing feathers

back: top of body, from the base of neck to the beginning of the tail

bantam: miniature poultry (formerly accepted as being one-fifth the weight of the large breed it represented, but nowadays one-fourth)

barring: alternate stripes of light and dark across a feather

beak: the two horny mandibles projecting from the front of a bird's face

bean: a black spot or mark, generally raised, at the tip of the upper mandible of a duck's bill

bill: a duck's beak

bow-legged: describes ducks that have a greater distance between legs at the hocks than at the legs and feet

breast: the front of a fowl's body, from the point of the keel bone to the base of the neck

breed: a group of birds breeding true to type, carriage, colouring and characteristics of the breed name they take

breeding flock: adult ducks used for reproduction purposes

brooder: a box or enclosure (usually warmed by a light globe or a heat lamp) in which the temperature is controlled and young birds are placed immediately after hatching

broody/broodies: used to describe birds that are ready to sit on a clutch of eggs

candle: the process of using light to determine the interior quality and stage of development within an egg

carriage: the bearing and attitude, or style, of a bird when walking

caruncles: the red, fleshy protuberances around the beak and eyes of Muscovy ducks and drakes

cloaca: the small opening under a duck's tail common to the digestive and urinary tracts from which body waste is emitted, and the reproductive tract through which eggs are laid

cobby: short, round or compact in build

collar: a white mark that almost encircles the neck of Mallard, Rouen and Duclair drakes (also known as the 'ring')

concave sweep: hollow curve from the shoulders to part way up the tail

condition: state of a bird's health; brightness of eyes and face, and freshness of plumage

coverts: covering feathers on tail and wings

crest: tuft of feathers on head, shaped like a half-moon

crop: a digestive tract storage organ

crossbred: a duck produced by crossing two or more different breeds or varieties

culling: the removal or disposal of inferior birds from the flock

cuticle: thin, skin-like layer covering the outside of an eggshell

dished bill: depression or hollow in the upper line of the bill of a duck or drake

down: initially, the soft fluffy feathers covering newly hatched ducklings; also the fluffy part of the feathers along the shaft and at the base of the feathers of adult birds

drake: a male of the duck family

duck: web-footed swimming birds of the genus *Arras*, characterised by a broad, flat bill and short legs; also the female of the duck family (*also see* 'hen')

duckling: a young duck of either sex, before all the down has been replaced with adult feathers

eclipse moult: occurs in the warmer weather after the breeding season when the bright plumage of the Mallard, Rouen and Duclair drake is temporarily replaced with plumage similar in colour to that of the duck

egg tooth: a temporary projection of the bill used for breaking out of the shell

embryo: a young organism in the early stages of development; a gosling, duckling or chicken before it hatches from the egg

face: the skin in front of, behind and around the ears

fertility: the capability of producing an embryo

flight coverts: small, stiff feathers covering the base of the primaries

flights: primary feather of the wings used to fly, but folded out of sight when the bird is resting

free-range: a poultry management system where poultry range outside their night houses during the day

green duckling: term used for young Pekin ducklings (bred for ease of meat processing) at 6–9 weeks, before the 'pin' or quill feathers of the primaries have developed

ground colour: main colour of body plumage on which markings are applied

gullet: the loose part of the lower mandible

hackles: the neck feathers of a fowl

hard feather: close, tight-knit feathers, as found on game birds

hatchability: the ability of eggs to hatch; can be expressed as 1% of the fertile eggs set (the total number of eggs, as opposed to eggs set x 100)

head: comprises skull, comb, face, eyes, bill, ears and wattles

hen: a female fowl after the first moult (female ducks can also be referred to as hens)

hind toe: the fourth or back toe of a fowl

hock: joint of the thigh with the shank, sometimes called the knee or elbow

incubation: the process by which young birds are hatched from eggs, including natural hatching by the mother and artificial hatching in a machine (incubator)

incubator: a box able to maintain a constant temperature of approximately 39.4°C for hatching eggs

imprinting: development of recognition and attraction in young birds

keel: blade of the breastbone (in ducks, the dependant flesh and skin below it)

keel bone: breastbone or sternum

lacing: a stripe or edging all around a feather, differing in colour from that of the ground colour

leg: the shank or scaly part

mandibles: horny upper parts of bill or beak

markings: the barring, pencilling or lacing of plumage

milking: collection, on a daily basis, of eggs for incubation

moult: the natural replacement of old feathers with new

outer lacing: lacing around the outer edges of a feather as opposed to inner

oviduct: the duck's sexual organ

pencil: the drake's sexual organ

pencilling: small markings or striping straight across a feather

pin-feathers: new feathers that are just emerging from the skin

pip: the first visible break in the shell made by a duckling, gosling, chicken, etc.

primaries: the outer 10 feathers of the wing used for flying, but tucked out of sight when the bird is on the ground

purebred: ducks, geese or fowls that are of a specific breed and that have not been crossed with a different breed

pupil: black centre of eye

set: eggs that are in the process of incubation

quill: hollow stem of feathers attaching them to the body (also known as 'shaft')

roach back: humped back, or arch in back

saddle: the posterior part of the back, reaching to the tail of the male, and corresponding to the cushion in a female

secondaries: the quill feathers of the wings, visible when the wings are folded and closed

serrations: 'saw tooth' sections of the mandible

sexing: the act of determining whether birds are male or female

shaft: the stem or quill part of the feathers

slipped wing: a wing in which the primary flight feathers hang below the secondaries when the wing is closed (also known as 'split' or 'angel' wing)

sport: an animal or plant that shows an unusual deviation from the normal or parent type – a mutation

strain: a family of birds from any breed or variety carefully bred over a number of years

tail feathers: straight and stiff feathers of the tail only

thigh: the part of the leg above the shank and covered with feathers

trio: a male and two females

twisted feather: the shaft of the feather is twisted out of shape

type: shape and mould of bird

under colour: colour seen when a bird is handled, that is, when the feathers are lifted; colour of fluff of feathers

variety: a definite branch of a breed known by its distinctive colouring and markings

vent: small external opening under the duck's tail through which eggs are laid

waterfowl: birds that spend most of their lives on or near water, including ducks, geese and swans

wattles: the fleshy appendages at each side of the base of the beak

web: a flat, thin skin that joins the toes of waterfowl

wing bar: any line of dark colour across the middle of the wing

wing bow: the shoulder part of the wing

wing butt: the end of the primaries; the ends of the wing

wing coverts: the feathers covering the roots of the secondary quills

wry tail: describes a tail carried awry, to the right or left side of the continuation of the backbone

Bibliography

Agnote (no date) *Meat Ducks in Western Australia.* Department of Agriculture, Western Australia.

Australian Quarantine and Inspection Service (no date) *Import and Export Conditions.* <www.aquis.gov.au>.

Holderread D (1978) *Raising the Home Duck Flock.* Garden Way, Yermont, USA.

The Poultry Club of Great Britain (1960) *British Poultry Standards.* Poultry World, Great Britain.

West G (Ed.) (1988) *Black's Veterinary Dictionary.* 16th edn. MRCYS, London.

www.ingramcontent.com/pod-product-compliance
Lightning Source LLC
LaVergne TN
LVHW060623110826
845147LV00015B/921

9780643106512